FACILITIES PLANNING &RELOCATION

David D. Owen

FACILITIES PLANNING &RELOCATION

A complete management system including:
- *Comprehensive text*
- *Sample specifications*
- *Over 50 forms...*
  reproducible and on disk

David D. Owen

R.S. MEANS COMPANY, INC.
A Southam Company

CONSTRUCTION PUBLISHERS & CONSULTANTS
100 Construction Plaza
P.O. Box 800
Kingston, MA 02364-0800
(617) 585-7880

The editors for this book were Mary Greene and Janet Pogue. Production was managed by Helen Marcella, and coordinated by Marion E. Schofield. Composition was supervised by Joan C. Marshman. The book and jacket were designed by Norman R. Forgit.

Illustrations by S. Michael LaNasa.

10   9   8   7   6   5   4   3   2   1

Library of Congress Cataloging in Publication Data

ISBN 0-87629-281-3

TABLE OF
CONTENTS

Chapter 3. Programming to Determine Actual Space Needs

Chapter 4. Scheduling and Budgeting the Project

Chapter 5. Schematic Design

Chapter 18. Color, Paint and Wall Coverings

FOREWORD

The radical restructuring of American and international business operations is moving at an unprecedented pace. "Right-sizing," mergers, acquisitions and spin-offs are the norm. The cost of reshuffling real estate assets and corporate resources is in the billions of dollars. The pressure is on to be a low-cost operator and to maximize the value of facilities capital expenditures. Corporate facility managers must produce more responsive and proactive facilities to support both long- and short-term business needs. At the same time, they must minimize cost and employee disruption, as well as maximize productivity by providing quality and efficient work environments.

Facility management, as a formal profession, is not very old. As such, there are currently few books specifically geared to the facility manager's or planner's role in leading relocation projects. Planning and managing a corporate facilities relocation is complex and demanding. Strategies for people, technology, leases, furniture, construction and space must be developed and monitored effectively to meet schedule and budget constraints. This book addresses the various tasks, and outlines the roles and responsibilities, as well as how and when the activities should occur. These strategies focus not only on managing the design team, construction and internal clients, but also the pre-design activities that should occur prior to beginning an actual project. The pre-design tasks include deciding if a relocation is necessary; establishing a team, budget and schedule; evaluating space options; and selecting the right building and programming.

David Owen draws on his many years of experience in sharing his personal insights on how a corporate relocation can occur smoothly. He has written this book to lead the inexperienced as well as the experienced facility manager through a complete project. Owen presents the information sequentially and provides numerous checklists, forms, surveys and samples for use during a project.

Furthermore, he provides such basics as technical background information on flooring, lighting, paint, plantscaping and artwork, to assist the various architects, designers and other consultants on the project team. Sample letters and outline specifications are also provided for facility or project managers who prefer to coordinate these items in-house.

There are numerous decisions to be made during the course of every relocation project, but rarely is there time to thoroughly investigate and implement each issue once the project is under way. This book is an important reference tool to help outline which decisions need to be made, and when, and the impact of each decision. The book allows readers to leverage Owen's experience, so the effect on the facilities department, the firm's clients, and its bottom line will all be a grand success.

Janet A. Pogue
Senior Associate, Gensler & Associates Architects

ACKNOWLEDGMENTS

As with all efforts, no one does anything truly alone. This book owes its existence to a number of individuals who contributed their skills and knowledge to it in one form or another.

I must first acknowledge the special contribution of M. Arthur Gensler Jr., FAIA, of Gensler and Associates Architects, who in 1977, personally sat me in front of a chalkboard and outlined his concepts of space programming and planning. A special acknowledgment to Edward C. Friedrichs, AIA, as well, also of Gensler and Associates Architects, who a few years later helped me to brainstorm the outline and contents of the book and assisted in the preparation of the original format.

The review by Robert D. Vrancken, PhD, Director, Facilities Management Program at Grand Valley State University, encouraged me to complete the revisions to the original text and to complete the work.

My thanks to a highly qualified group of advisors who listened to my concept for this work and then reviewed sections and provided me with their comments: Henry G. Williams, FIES, Senior Lighting Specialist, General Electric Company; J. Marshall Hemphill, Research Manager and G. Robert Spalding, Senior Research Scientist, both of Armstrong World Industries; to Debra Kates, Manager of Horticulture Services, Plant Design Group, Inc.; and to Diane Stegmeir, President, Corporate Art Concepts.

A special thanks to Janet Pogue, Gensler and Associates Architects, who provided me with fresh input and editorial advice, and who made this a far better work.

Finally, my special thanks to those who provided computer assistance: Sweets Electronic Publishing for their CADD work, and a very special thanks to S. Michael LaNasa who, with his entire family, worked on the structure of this book, far beyond their reward.

INTRODUCTION

It was not until the early 1960s that the first speculative office buildings were developed in the United States. These buildings opened the door to new forms of office usage. Through World War II and for a decade thereafter, U.S. industry conducted business essentially as it had since the great industrial expansion triggered by the Civil War. A company would bring a basic material, product or service to the market, whether it was steel, an automobile, insurance, banking or a telephone. Once introduced, the product would undergo improvement changes, but rarely was it subjected to explosive technological changes or rapid design and market obsolescence. Beginning in the 1960's, U.S. industry found that products it had been producing and marketing for decades, in essentially the same format, had to be redesigned or replaced with new or revised products. Technology and competition created a new industrial environment. Changes to redesign or replace required the addition of new types of personnel; market and product research, national, regional and local sales groups, project control divisions, etc., often functioning far from the "plant" or "home office." The new offices were also under the stress of change, growth and relocation, usually with the addition of whole new departments dealing with evolving business concepts, equipment and practices.

The construction industry was also moving along a parallel track. Technological advances in construction materials and systems began to have a major impact on the design, construction techniques and costs of new buildings. The evolution of central air-conditioning systems, curtain wall construction and higher speed elevators were among the developments that led to major revisions in office building designs and, in particular, in building floor plate sizes, building heights and floor-to-floor heights. The new, expanded floor plates generated new core-wall to exterior glass-line distances and changed the way floor space utilization was viewed. The development and acceptance of systems furniture, suspended (plenum) ceilings, and fluorescent lighting troffers, and the use of carpeting instead of floor tile propelled space utilization toward revolutionary solutions.

With the creation of these new buildings, a small group of architectural firms began to experiment with the concept of space programming: the analysis and definition of specific workstation and department requirements, and circulation around and to them. Special areas were being specified and added, such as conference and training rooms, food service, storage and copy centers. A recognition grew that all of these many space needs had to be defined, assessed and documented, and each space evaluated as to its specific anticipated growth needs. What became apparent to these architects was the need for a process, a system, a program, with standards and appropriate definitions of today's requirements by which their clients could forecast tomorrow's needs. Through experimentation and a great deal of creative work, a few leaders in the interior design community began to develop the concept of defining space needs by converting individual workers' space requirements into standardized *workstations*, that went beyond mere desks. These workstation definitions were grouped into functioning departments. The departments were placed in adjacency patterns for efficient communication with other departments and facility elements, with each department having its unique forecast of present and future space needs. The evolving results were programs that translated into logical and responsive space plans, design and construction documents. These assisted the client in changing and growing without major department interruption or dislocation.

While there are now a sizable number of architectural and design firms capable of providing quality space planning, it is not always clear to the user who is really qualified and who is not. Even the most qualified design firms can produce a quality product for their client only if the client has a clear understanding of its own role in the process, what information must be conveyed to the planner, and how the planner should use that information.

This book will prepare the user for effective participation in the space planning and development process by explaining how the process works, including some of the planning and design conflicts that inevitably arise. Most importantly, the book provides the user with easy-to-follow guidelines for dealing with the process and any conflicts. The book is an attempt to pull together various processes, formats, and material which have been used over and over in producing successful office facilities. We have tried to direct the user in establishing and maintaining controls over the project through the use of detailed programs, criteria, studies, budgets, and critical dates schedules, together with information on some of the more important systems, materials and furnishings to be specified and purchased. The user may find differences between the processes contained in this book and some actual experiences, but all of the critical functions described herein will have to be dealt with.

It is logical and economically desirable to use as little space as possible for a facility. It is also logical and desirable to develop a facility which will permit reasonable expansion in the specific areas where expansion is anticipated, rather than merely taking an option on a piece of space next door and planning on expanding into it... someday. This latter form of expansion is a "growing like Topsy" process which will result in poor adjacencies, poor department organization, poor communication and a great deal of unnecessary department reshuffling at great long term expense.

Even though many architectural firms today are very good at office facilities planning, one should not assume that consultants are totally knowledgeable or fully understand the needs of a facility. Facility managers must do their own research, and ask questions. The designer does not know how to conduct your business. You do. This is to be your facility, not the designer's. There are hundreds if not thousands of decisions which must be made by someone. Often, some of those decisions are made by people who are not qualified. Mistakes or omissions go unrecognized because they were papered over with corrective solutions or went completely unrecognized out of ignorance, until well after move-in, when corrective action is both costly and disruptive, assuming a correction is ever made.

Developing a major office facility is a major undertaking, by-in-large thankless. If it goes well, everyone will take the results for granted, with the designer probably receiving the credit. If the job does not go well, the responsible coordinator or facilities manager will take the heat.

This book is designed to help facilities professionals to better understand some of the questions to be asked and to prepare them to develop sound answers. Use the book as a tool. Add to it. Modify its contents to suit your particular and changing needs. Our pride of authorship will be in your ability to understand and use the processes; to modify and to improve upon its contents.

Part I

THE PLANNING AND CONTROL PROCESS

This book has been organized in two parts. Part I deals with the processes and concerns of determining if a relocation is necessary or desirable; researching for and acquiring a new facility; programming, planning and preparing for a relocation; and finally, executing the move itself.

Because of the diverse readers whom this book may address, we have placed the reader in the general role of a facilities or project manager within a "Company" or other type of institution. The subject matter is presented in sufficient detail to provide the user with the necessary tools to deal with the various professionals and concerns that may be encountered. The material has been presented in the same sequence as the reader might find it during an actual planning and relocation process.

The procedures and solutions presented are by no means the only ones that will do the job; however, they have worked well for us.

Chapter 1

GETTING STARTED

Introduction   Many functions have been assigned to office space, but from our perspective, it has but one primary function, and that is to provide an efficient location and environment in which to produce work. Whether this work-product takes the form of legal consultation, a medical practice, a banking function, or the control center for a manufacturer, the office space is, in the final analysis, a work-product environment. We should approach the selection of office space, as well as its planning, design, furnishing and occupancy, from the primary point of view of an industrial engineer, taking as our main concerns the *function, cost,* and *image* of the facility.

Function

An office should have a space-utilization efficiency near to that of a high-tech manufacturing facility. An office is a place where people and information sources are brought together, and in which they can communicate with one another and with other people outside of the office. An office is also a place where people research, manipulate, team together, create and document information. It is therefore a place for people to work. It is important that the total environment in which people work be as carefully designed and equipped as possible in order to maximize their efficiency and accuracy. It should be located in an area easily accessible to its labor force, its clients, and its public. It should be studied and designed with careful consideration to work flow, interrelationships between functions, visitors' needs, employees' needs, and changing space requirements. Attention to these design needs and your knowledge and skill in solving your company's unique and constantly changing methods of doing business will determine how efficiently and effectively your work space will function.

Image

Every office portrays an image. The location and design of an office projects an image. Each company should carefully examine how it wishes to be seen by the various elements of its universe: its customers, its peers, the public, the community, and its employees. More often than not, the matter of image is left in the hands of the public relations or advertising departments. However, many companies are now examining their total environment from the point of view of image. They are asking the question, "Does our space and its location project the image we wish to portray to the various groups viewing us?"

Most companies attempt to project a certain image to their clients, and often a different one to the public. In many instances, the matter of image can be important to intra-office relations or marketing. Or sometimes the concern is the image portrayed to employees. Image is principally subjective and is often closely tied to a company's sense of its own culture. There is nothing wrong in developing an office environment with a strong emphasis on image, nor is a high image design necessarily more costly to produce. The image desired by one company may be one of thrift and may be designed to reflect that intent. Most assuredly, certain types of offices should display a character that relates to their work-product. However, in designing new offices, be certain to differentiate the elements and costs ascribable to pure image from those belonging to basic occupancy and function needs. Designing an office space with a strong, well-conceived image need not carry with it a high cost.

Cost

Every person and most activities within an office are cost or expense centers. These costs can be broken down into three major categories:

- **Direct Occupancy Costs:** rent; and landlord surcharges, such as electricity, common area maintenance, parking, amortized landlord-provided tenant improvements, etc.
- **Amortized Costs:** furniture, partitions and dividers, non-landlord-provided improvements (such as special lighting, lunchroom facilities, special wall covering, etc.), items paid for by the tenant.
- **Direct Labor Costs:** Salaries, benefits, perquisites, etc.

Example

In order to get a feel for the scale of direct occupancy and amortized costs, consider the following example in Figure 1.1.

Based on an Average of 250 SF/Person:

Direct Occupancy Costs/Person		$/SF/Year	$/Year
Rent		20.00	
Utilities		2.00	
		$22.00	$5,500.00
Amortized Costs/SF	$/SF		
Interior design @			
$.75 to $6.50	4.00		
Tenant improvements by Tenant @			
$10.00 to $80.00	20.00		
Furniture and equipment @			
$15.00 to $50.00	35.00		
Moving	2.00		
Telephone @			
$500.00 per station	2.00		
	$63.00		
($63.00 X 250 SF) =	$15,750.00		
Amortized over 5 years @ 10%		$16.60	$4,150.00
Totals		$38.60	$9,650.00

Figure 1.1

This example represents a very plausible annual overhead item for each workstation ... whether occupied or not. One of the most critical elements is that the $15,750 was spent up-front, and can only be recovered against future income.

Until recently, it was considered sufficient when designing a new facility to simply calculate future personnel growth to the date of move-in, plus an additional five years forward, and use these figures in determining a company's space requirements. Today one must consider a whole new set of factors in the planning process: such changes as computer technology, especially the personal computer (PC), shrinking data processing space needs, centralization or decentralization of functions, and evolving personnel needs. These other factors make it important to be able to reconfigure existing facilities to adapt to these changing needs. The office work environment should be able to adapt without causing economic hardships or slowdowns in work output.

Deciding If a Change Is Necessary

A move, an expansion, or a remodeling program is under consideration. Is it really necessary? One way to find out is by preparing (or having prepared) a written report answering the questions listed below. The responses should be based on a tour of the present facility and discussions with department and section heads. The questions are:
- What are the positives and negatives associated with the current office space?
- Are there special requirements which cannot be accommodated in the present space, such as special security needs, special electrical requirements, frequency and volume of special pick-up and delivery needs, special storage requirements, etc.?
- What is the history of growth of each department in numbers and percentages over the past five years?
- What is the anticipated personnel and square foot growth or shrinkage of each department in numbers and percentages over the next three and five years? (staff counts and square feet).
- What planned or anticipated organizational changes will affect the long-range space needs of the facility (e.g., technological changes, adding or deleting departments, subsidiary companies, etc.)?
- Is there enough space in the present facility to handle these needs, either in its present form or after remodeling?
- In the opinion of the person(s) writing the report, which of the following alternatives is advised at this time?

  ( )no change   ( )move   ( )expand   ( )remodel   ( )contract

Estimating Preliminary Square Footage

If the responses indicate that a change is warranted, it is necessary to estimate how much and what type of space will be needed; then it is possible to determine whether it's best to move, expand, or remodel.

Once approval has been given for further planning, it will be necessary to make preliminary estimates of the amount of space that will need to be leased or remodeled. Because moving and expanding are expensive and time-consuming, it is important to lease or remodel enough space to serve the organization's needs five to ten years after the "change date," — the date when the remodeling work or move-in to a new facility is complete. By providing a growth or reduction strategy now, you save time and money over the long run.

To estimate the total amount of needed space, estimate the total number of employees the facility will require three and five years from the change date. To do this:

- Decide the approximate date by which the change should be completed.
- Ask all department heads to estimate their respective department's total number of employees five years after the change date. (They can do this by evaluating their present personnel, then estimating the total number of employees they will have as of the change date, and finally, using projected business volume and related staffing requirements, estimate their total personnel three and five years after the change date.)
- Total the projections for each department.
- Add in or delete totals for any affiliates or subsidiaries that will be added or removed from the anticipated facility.

Caution: At this point, the estimates should be reviewed, in detail, by senior management. Frequently, there are strategic plans involving groups or departments which have not been announced and which may substantially affect your analysis.

This total represents the estimated number of employees the facility will require three and five years after the change date. To cross-check the estimated total, compare the predicted percentage increase in personnel against your company's current annual growth rates (sales, product lines, services, departments, personnel, etc.). The following is a fairly accurate approach for preparing a quick estimate of how much space might be needed for the new, or remodeled space.

Density Range

The quickest method is the use of density ranges. These ranges were developed from surveys of large office space users. The term *density* refers to the average number of usable square feet of space per person occupying the space. A series of recent surveys conducted by Gensler and Associates indicated that the average density was approximately 224 USF (usable square feet) per person. The density ranges disclosed from these same surveys were from 90 USF/person to 350 USF/person. ["Usable Square Feet" is a specific term as defined by the Building Owners and Managers Association (BOMA).]

For a preliminary estimate of the company's existing density, take a look at the present facility. Determine the number of people occupying it. Determine the number of usable square feet (USF) that is being occupied or used. Divide the USF by the number of people. Now consider the type of space utilization the company is currently using.

If the company is presently utilizing mostly private or enclosed offices for most of its employees, a slightly larger density factor and circulation factor will be appropriate than if the company uses mostly open plan workstations. For the purposes of this estimate, we suggest using about 28% of the usable square footage as a circulation factor. This number is really made up of two different circulation factors: a 16% factor for getting around between workstations, offices and support facilities contained within a given department, plus a 12% factor to get between one department and another. Both factors are approximations and will vary from one facility to another. If a facility is oriented towards more private offices rather than an open plan, it may be necessary to increase both factors by two to three percentage points.

The next step is to consider whether the present space utilization philosophy is still appropriate. Are there trends or changes in the manner in which the company does business that would shift the present space utilization policy toward more open plans or toward a greater use of enclosed spaces?

Depending on the company's perceived philosophy toward current and future space use, it should be possible to make a quick density assessment. It is likely that the average square footage of existing usable space per person will fall on the low side of 200. If the 200 square feet per person currently available feels cramped for space, we suggest using a range of 225 to 250 square feet per person as the new usable requirement. Since the cost of office technology is decreasing, the current trend is for slightly larger work spaces per person to accommodate more individual equipment and a trend away from shared computer and computer-related equipment.

Reorganization and Contraction

Until the latter part of the 1980's and the early 1990's, usual space planning evolved around a problem of providing for orderly and anticipated personnel growth. Two or three trends began to influence the dynamics of office area requirements. These requirements began to change much more rapidly than we had ever experienced, and dealing with them has required new techniques. One day we are planning for a five-year growth pattern. A few months later, we are required to reorganize large portions of a space due to a new business approach, the influx of new employees, or the transferring out of entire departments. These ensuing influences began to become a major trend through the merger or acquisition of companies and the reorganization of major departments. A facilities manager may, at one time, be dealing with the problem of expanding space for a newly acquired marketing department in one facility; at the same time, the same facilities manager may be eliminating occupancy of an entire floor or building as a result of a transfer of a major accounting or telemarketing group to another city.

A second trend is a result of sophistication, increase in power, and a decrease in the number of people required. PCs, "smart" terminals, fax transmission, electronic mail or giant databases have increased the productive power of individual employees to the extent that one person can often comfortably perform the tasks of three to four people just five to six years ago. Not only have people become more productive, but the equipment they are using has shrunk in size so rapidly (and is still shrinking), that no additional workstation space may be required to accommodate this new high-productivity employee. There seems to be an apparent contradiction in the growth pattern of individual workstations. On the one hand, many workstations are being enlarged to accommodate more equipment per employee. On the other hand, the size of the equipment is decreasing.

The final trend is the constantly reoccurring streamlining of businesses in order to achieve economic efficiency. Most businesses are now seeking ways to reduce operating and administrative expenses. Increasing efficiency or reducing office staff will result in needed modifications to office space either through reorganization or contraction. As a result of these newer dynamics, our knowledge of space planning must also encompass reorganization and contraction, along with growth.

If the results of the estimate indicate a reorganization that will result in a major remodeling, it may be necessary to begin to think in terms of contraction of space. Your planning and recommendations in this regard must now take a number of difficult conditions and corporate policy considerations into account, as the space planning program is developed.

Assuming a reorganization of existing space is indicated, rather than growth and relocation, a series of critical questions must be reviewed and the answers agreed upon before proceeding with any further steps. Often the direction of a strategy will become evident as answers are developed for these questions.

- How many people will remain in this facility?
- How many people will require private offices, as compared to the number of private offices currently available?
- Will the remaining private offices be in the right locations and of appropriate grade standards to properly serve the reorganized facility?
- If the grade standards are inappropriate, is the Officer in Charge willing to "functionally override" the variances from current policy standards? If not, more demolition and rebuilding may have to be considered.

After a review of the reorganized private office needs and all other "open plan" personnel workstations, the same types of questions must be asked regarding all of the common elements, including:

- Reception Areas
- Conference Rooms
- Education and Training Rooms
- Lunch Rooms
- Copy, Fax, Printer Centers
- Libraries
- Computer Rooms
- Data Entry Rooms or Spaces, etc.

Now ask the questions in reverse. How many of the common elements may have to be added or substantially modified for the reorganized space?

As the space evaluation process progresses, the presence of existing spaces which have been designated to remain will complicate the adjacency studies and hierarchies. If there are rooms, spaces and facilities to remain, they should be marked as *Common Elements* on the adjacency plans in some unique manner so as to indicate their relationship to the added or relocated departments.

Space Reduction

When the evaluations are complete, it may be determined that a reorganization is necessary, and that the result will be a permanent or long term (over five years) actual reduction of space needs. The question is, what to do with the unneeded space.

Subleasing

Again, we are faced with a number of questions which must be answered before we can proceed with a recommendation or attempt a solution.

- How much space will be freed up?
- Can the remaining personnel and functions be consolidated to free up an entire floor for lease or sublease?
- Will it be a single block or, at least, fairly large blocks of space?
- Will it be on one or multiple floors?
- Will the floors be occupied by one firm; or will they be multi-tenant or can they be converted to multi-tenant? Careful consideration should be given to converting a single tenant floor to multi-tenant floors. Building and Life Safety Codes for multi-floor space are very different from single floor occupancies. Access to toilet rooms and emergency exiting should be considered as well.

- Assuming the multi-floor questions can be resolved, it must be determined whether the lease with the facility landlord will permit the firm to sublease a portion of the space. See "Lease Terms" in Chapter 2.

So far, we have not addressed the really difficult questions. These are the ones dealing with the sublease market.

- Are there potential sub-tenants available to occupy the space?
- At what rental rates can the space be leased?
- What will the cost be to provide a shell space or tenant improvement budget for prospective tenant(s)?

Note: Shell space is a particularly thorny problem when dealing with heating and cooling systems, costs and controls, and with electrical distribution, development of electrical load centers, and metering. [Sub-metering may or may not be permitted under PUC (Public Utilities Commission) or the local utility's regulations, or under the terms of the facility lease.]

- What sort of shell space will be provided?
- What sort of "Tenant Improvements" package will be necessary and at what cost?
- Will a broker be required to lease the space?
- What will be the brokerage commission costs?
- What will be the legal costs in preparing and negotiating the sublease?
- What will be the relationship of the sub-tenant to the facility landlord—both legally and in terms of a working relationship?
- What will it cost to become a landlord, i.e., what will be the overhead cost to invoice a tenant (or tenants), collect and account for the rent?

Frequently, services due the firm from the landlord under the terms of the lease can be passed along to a sub-tenant.

Assuming that answers can be found to all of the above questions, there may be other difficult issues requiring difficult solutions. It would not be unusual to wind up with large amounts of space unaccounted for—possibly full floors. These kinds of problems may require outside help. Efforts to sublease may not be fruitful; the company's management may not want to be in the "landlord" business. Many other companies in the area may have the same kinds of problems and may be looking for the same kinds of solutions.

Lease Cancellations

Unless there is someone on staff with a real sense of the local leasing market (or sale market for those who are selling), and knowledge of what kinds of lease cancellation deals can be made, the best source of help is a top quality real estate broker who has the "street smarts" to help negotiate a lease cancellation with the facility landlord. Lease cancellations are simply a means of buying oneself out of a lease. Often a landlord will consider a lease cancellation deal. The following are some rules of thumb which can be used to determine whether this is a viable approach.

- If the lease is one year or less, the landlord is not likely to be interested in a buy-out at any discount.
- If two years or more remain, the company might offer one full year's rent, up-front, for the cancellation of the total remaining two years.
- For a lease in the five-year range, a fifty cents on the dollar deal is still a likely choice. In other words, the offer might be to pay, up-front, one-half of the monies owing on the outstanding term of the lease.
- For a lease longer than five years, the fifty cents on the dollar equation might be reduced to a lesser cost. This will require a keen knowledge of current market conditions. One of the most important elements in lease cancellation strategies is the desirability of the space that is being relinquished. There is no substitute for quality space in quality buildings in quality locations. None.

Project Management Teams

If the conclusions of the analysis indicate either the reduction of space or the addition of space, many more participants will have to become involved, some within the company, some as consultants, some in adversarial roles — such as landlords, with negotiations that will have to be conducted. Before beginning the divestiture or acquisition process, it is important to identify and involve the various teams and team members. They should be properly informed and have the authority to make the necessary corporate and technical decisions.

Two permanent management teams should be established. In order of hierarchy, the two teams might be titled *Project Review Team* and *Project Management Team*. The Project Management Team will be performing nearly all of the analytical functions and making a major number of the project-related decisions. The Project Review Team will function in a review capacity and have responsibility for determining project strategies and making the economic decisions. The Project Review Team must be able to authorize commitment to the next Phase or to terminate the project.

The Project Management Team

The Project Management Team deals with the day-to-day considerations. It generates or secures most of the data necessary to move forward through each step of the project and to continuously define the project, both physically and financially.

The Project Management Team members will likely consist of two types: the day-to-day operational people, and those who provide internal information, direction and suggestions. The day-to-day operational team members function as the professionals. If these functions are provided from within the company, the person or persons selected should have hands-on project management experience. It will fall upon these individuals to select, negotiate with, and direct the various professionals: architects, engineers and other consultants who will be involved in the project. They will be called upon to render countless evaluations and decisions every day of the project. They must be able to organize and coordinate all of the various activities of the consultants and contractors, while never losing sight of the policies, strategies and ultimate objectives of the company. Most importantly, they must anticipate the questions that are not asked, as well as answer the questions directed to them, both from within the company and from outside.

As the project progresses and more outside team members become involved, the so-called Project Team begins to take on the characteristics of a managed group. The various team members are yoked together for the purpose of producing the project. Never lose sight of the fact that each of the team members has his or her group's primary interests in mind, which will take precedence over the overall objectives of the project. All team members who are outside contractors are, from a legal standpoint, in an adversarial relationship with the company, even though they are attempting to work in unison with the company and the other team members. A good project manager must recognize the self-interests and should be able to satisfy them, while maintaining the goals of the project. It is obvious that a cooperative team is most desirable. To accomplish cooperation, there must be strong and respected leadership. It is the function of the Project Management Team's Project Manager to supply this leadership. The alternative may be a battle of egos and finger-pointing sessions.

Another component of the Project Management Team consists of those representing the specific company interests. The key member of this group should act as a Primary Liaison, and should be a fairly high-level employee of the company. This person should have a reasonably high degree of project and real estate knowledge, or at least have an engineering- or facilities-related background. He or she should be capable of understanding the technicalities of building systems and problems. This individual might be the facilities manager, and should have reasonable access to all levels of authority within the organization, at least high enough to obtain project approvals and get checks signed. This person must be able to devote a sizeable portion of any day to project matters, making the daily decisions that will keep the project moving without delays. This is a key position, and should be filled with great care. It is possible for the principal Project Manager to serve this function.

A third member of the Project Management Team should be a representative of the Human Resources, Operations, or Executive/Administrative group. Because much of the project language may be foreign to this person, he or she should be a good team player, capable of moving up and down within the corporate hierarchy to gather information and secure philosophical and policy direction. The Human Resources member has the responsibility to see that the project produces the appropriate physical facilities for all the functions being performed within it, as well as providing the maximum possible benefits in the form of the working environment for the workers – from the Chairman of the Board to the mail clerks. Decisions must be made on dining facilities – from private, to cafeteria, to vending machines. Health care facilities, exercise facilities, child care, training, audio visual, copy centers, security, emergency response, visitor control facilities and the like must also be considered and programmed. The Human Resources member will be the one called upon, together with the key liaison person, to dig out the company's "would like" versus "must have" policies.

The fourth member of the Project Management Team is a representative of the financial side of the company, either the Comptroller's or Treasurer's Departments. During the preliminary stage of the project, when economic considerations are being reviewed, including alternative costs and expense budgets or pro formas, it is most important that someone who has the knowledge and ability to speak from the financial side of the company be available for regular discussion and response. Once the project is committed, this person may have a declining role to play. Most of his or her decisions will have been made, and if these were based on valid assumptions, little or no new evaluations will have to be made. However, this person will bear the primary responsibility for following the budget performances, for audit and for securing the necessary funds to satisfy the normal monthly progress payments.

The final member of the Project Management Team may not always be necessary. This person would represent "other assets." If the proposed facility should house people or assets which are not strictly office function-related, the team should have a member qualified to speak on behalf of those assets and needs. Such concerns as vehicles, communication, computer and processing centers, special screening or sound rooms, research, testing and development laboratories or shops, production and warehousing may be a part of this project. These assets and functions should be represented by someone who has the appropriate knowledge. As with the other members of the team, some daily response and effort will probably be required of this individual throughout the life of the project.

The Management Review Team
The Management Review Team should consist of the Project Liaison and, on an ad hoc basis, the Project Manager, together with two or three senior management people.

Depending on the significance of the project, the senior management members may consist of the Branch or Regional Office Supervisor, or even the Chief Executive Officer (or that person within the company who can authorize the commitment to the project from phase to phase). It is possible that this person will be securing authority from the Board of Directors. Consequently, he or she must either be able to make the appropriate presentation or have the venue for doing so.

The second senior member should be either the Comptroller or Treasurer, or some member of the Finance Department capable of locating and committing the funds necessary to cover the entire project and of assessing the various effects of the project on the company's balance sheet and earnings statement.

A third senior member may be the in-house counsel or the General Counsel, if any. If there is no in-house counsel, the review team might consider inviting outside counsel to the review meetings, not only to act as recording corporate secretary, but also to advise the other members of the legal consequences of their various commitments and their authority to act without the direct approval of the Board of Directors.

The various team members have been defined as a result of their unique and necessary capacities to contribute either direction, liaison efforts, information, counsel, or authority to the project. It is vitally important that they perform their jobs in a timely and positive manner. There are more bad projects as a result of uninformed or uninterested management than as a result of bad professionals.

Roles and Responsibilities

It is important while moving through each of the processes described in this book to seek out the wide variety of skilled assistance available. No commitment, financial or otherwise, should be made without assessment by either the appropriate team members of the company's in-house staff or consultants who are qualified in fields related to space planning and facilities management. This book also outlines a number of approval procedures.

To avoid confusion in these matters and to establish consistent standards of responsibilities, it is suggested that you develop a chart similar to Figures 1.2 and 1.3. This chart has been prepared to cover a fairly complex corporate structure. It can be revised or simplified as necessary. In any case, do prepare one and have it approved at all levels before proceeding. This chart is a device that can be used to secure the approval of the various management levels. It will communicate to the many individuals and groups their authority; and as the programming, planning and design steps are undertaken, such a chart can help to prevent a great deal of confusion and possibly some political turmoil. Use it to determine where to go for help and from whom necessary approvals can be obtained at each stage. (The chart also appears full-size in the Master Forms Section of this book.) It can be reproduced for easy reference and potential revision to reflect the company's unique structure.

In preparing a company-specific chart, be sure to determine what the company's policy is in regard to the tasks and decisions shown in the first column. The chart, as shown, indicates a strong corporate control. Some companies may prefer to merely set guidelines, and permit a great deal of regional or local control. The chart should fit the actual corporate culture and policy. If the company does not have a policy, try to have one developed and prepare the chart accordingly.

Roles and Responsibilities

Facility: _____ Location: _____

Prepared By: _____ Date: _____

		Branch		Region		Home Office		Professional Staff
Task/Decision		Department	Manager	Manager	Administration	Facilities	Corporate Facilities	
Establish Require-ments	1. Identify need for major change to facility							
	2. Gather data on present and future facilities requirements							
	3. Interpret data							
Locate Facilities	1. Site search							
	2. Evaluate alternatives							
	Financial							
	Physical							
	3. Finalize lease							
	Work letter							
	Legal							
Prepare Facility	1. Space plan and design							
	2. Construction documents							
	3. Select and coordinate contractors and vendors							
	4. Select and coordinate telephone and data systems							
Move	1. Select mover							
	2. Coordinate move							
Space Maint.	1. Cleaning and servicing							
	2. Fine tuning							

Figure 1.2

Roles and Responsibilities

Facility: __Branch__ Location: __Fairview, IL__

Prepared By: __David Johns__ Date: __11/22/93__

Task/Decision	Branch Department	Branch Manager	Region Manager	Region Administration	Home Office Facilities	Home Office Corporate Facilities	Professional Staff
Establish Requirements							
1. Identify need for major change to facility	Recommend	Access + recommend	Access + recommend	Review + approve	Access + advise	Access + advise	
2. Gather data on present and future facilities requirements	Execute	monitor + recommend	Review + approve	Review + approve	monitor	monitor	Monitor
3. Interpret data		Access + recommend	Review + approve	Review + approve	Execute	Policy	Advise
Locate Facilities							
1. Site search		Advise	Monitor	Review + approve	Execute	Policy	Advise
2. Evaluate alternatives		Access + recommend	Review + approve	Review + approve	Access + recommend	Access + recommend	Advise
Financial		Access + recommend	Review + approve	Review + approve	Execute + recommend	Access + recommend	Advise
Physical		Access + recommend	Review + approve	Review + approve	Execute + recommend	Access + recommend	Advise
3. Finalize lease		Access + recommend	Review + approve	Review + approve	Negotiate + recommend	Review + approve	Advise
Work letter		Access + recommend	Review + approve	Review + approve	Negotiate + recommend	Review + approve	Advise
Legal	Provide requirements					Review + approve	Advise
Prepare Facility							
1. Space plan and design	Provide requirements		Review + approve	Review + approve	Access, recommend Administer	Policy	Advise
2. Construction documents		Access + recommend			Access, recommend administer	Policy, review + approve	Advise
3. Select and coordinate contractors and vendors		Access + recommend			Access, recommend administer	Policy, review + approve	Advise
4. Select and coordinate telephone and data systems	Provide requirements	Review + approve		Execute by corp. communications	Coordinate with all parties		
Move							
1. Select mover		Access + recommend			Review + approve	Policy	Advise
2. Coordinate move					Access, recommend administer	Policy	
Space Maint.							
1. Cleaning and servicing	Monitor	Administer			Monitor	Policy	
2. Fine tuning	Identify need	Review + recommend	Monitor		Execute	Monitor	

Figure 1.3

14

Preliminary Budget Planning

To accomplish the first portion of this phase, a schematic version of the project must be defined. This can be done on your "desktop," without employing a lot of consultants or letting costly contracts. Once a simple paper or "desktop" version of the project has been defined, the various costs can be estimated, even though they will be very rough. From these rough estimates, a preliminary project budget and evaluation can be started. Out of these first organizational, preliminary project definitions and budgets, the management teams can make their evaluations and decisions as to whether to proceed to the next phase, or to reassess.

Before the project can go forward, preliminary estimates must be made of the total project cost. This figure varies widely depending not only on inflation, but also on the particular scope of the project. For example, the cost per square foot will be far different for making a small addition to existing space and using already-owned furniture, as compared to a major move to a new facility with all new furniture. While dollar amounts vary, the percentage allocation of each dollar should remain relatively constant. In the case of a major move to a new facility, involving the purchase of completely new furniture and equipment, the percentage breakdown of the construction dollar may be estimated as follows in Figure 1.4.

Once a dollars-per-square-foot figure has been developed, simply multiply it by the estimated total number of rentable square feet the project will require.

In addition, the changes in total annual operating costs (additional rent, additional employees, etc.) must be estimated. Refer to Chapter 4, "Scheduling and Budgeting the Project," for information on how to begin developing a detailed budget estimate.

Breakdown of Costs for a Typical Open Plan Office Facility Using All New Furnishings

General Construction	17.1%
Consultant Fees	4.5%
Directional Signage/Graphics	1.5%
Art Program	1.5%
Relocation Costs	3.9%
Unallocated Project Cost	8.0%
Furnishings	63.5% *

* If a large amount of existing furnishings or reconditioned or low-end furnishings are used, the shift will be between Furnishings and General Construction.

Figure 1.4

Preparing a Preliminary Schedule

Before opening lease negotiations for a new facility or for an addition to present space, one must have an idea of how long it will take after signing the lease until the new space can be occupied. When a landlord has completed a space according to the terms of the agreement, the company must start paying rent on that space—even if it is not ready to move in. To avoid paying rent on space that is not being used, be careful to accurately forecast the amount of time required to accomplish the necessary work prior to occupancy.

For facilities under 20,000 SF, consult the selected expert (e.g., architect or contractor) for advice on how to develop a schedule. For facilities over 20,000 SF, the graph in Figure 1.5 will be helpful. In developing a schedule, keep in mind that it will vary depending on the design complexity, the amount of special millwork and construction details, and specialty furniture and materials specified. See Figure 1.6 for a sample phasing schedule.

The graph helps to approximate (in weeks) the minimum time necessary before move-in. The time spans shown by the graph are only valid after lease negotiations with a specific landlord or agent have been substantially completed (that is, all major topics have been settled, and only small details

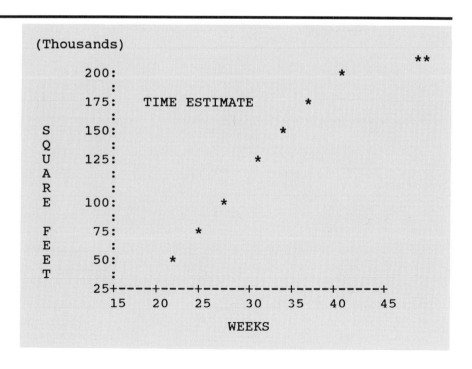

Figure 1.5

Phasing Schedule	Set Benchmark Dates in Space Below																							
Preliminary	50,000 SF																							
Weeks (Add additional periods as necessary)	1	2	3	4	5	6	7	8	9	10	11	12	13	14	15	16	17	18	19	20	21	22	23	24
Commitment by Client Company to Proceed	*																							
Preliminary Space Plans	***	***																						
Final Space Plans, Review & Revise			***																					
Design		***	***																					
Review and Approval				**																				
Construction Drawings	*	***	***	***	***																			
Coordination					**	**																		
Review and Approval						**	**																	
Furniture & Furnishings Plans & Specs			***	***	***																			
Review and Approval							**																	
Construction Bidding						***	***	***	***															
Construction 1st. Week Mobilization									***	***	***	***	***	***	***	***	***	***	***	***	:::			
Furniture & Furnishings Bid and Order					**	***	:::	:::	:::	:::	:::	:::	:::	:::	:::	:::	:::	...	...					
Furniture Installation																					***	***		
Move In																						*	***	

Figure 1.6

remain). It also assumes that programming has been completed to estimate the required amount of square feet. Programming may need to be reviewed, in detail, prior to beginning the design process. The vertical axis represents the size of the job in rentable square feet. The horizontal axis represents time in weeks.

Note that this graph shows the minimum times required and assumes prompt review and approval at all phases. Any unusual delays once the process has started will immediately invalidate this chart.

It is, of course, in a landlord's best financial interest to have the tenant begin paying rent as soon as possible. Some landlords may suggest the minimum times shown on this graph are too generous. It is important to know the company's own requirements, and refuse to be swayed.

A Word of Warning

While the project is being planned and various phases are being carried out, life will go on. Some things will go wrong, some things will go right — *most of all, things will change*. Be aware of the necessity to constantly review all of the basic information on which ongoing judgments are based. This includes schedules, budgets, and personnel projections. Keeping an eye on everything is the best way to ensure that the project will go smoothly.

At this point in the project, the lines of authority to be employed for the project should have been determined. The design image the company wishes the new facility to convey should be established. The amount of space of the new or redesigned facility should have been estimated, and defined in usable square feet. Project teams will have been designated, and a preliminary project schedule will have been produced. All of this is preparation for the next step in the project, that of examining options for new or redesigned space.

Chapter 2

EVALUATING SPACE OPTIONS

This chapter deals with the actual selection of office space, comparisons to other available space, negotiations with landlords, letters of intent, tenant work letters, renovations, expansion leases, and future expansions.

Renovate,
Add More Space,
or Relocate

If it is concluded that expansion may be necessary, look around the present facility. What is the net usable area presently under lease? What additional square footage is available for lease in adjacent space? If the sum of these two numbers is significantly less than the number of square feet the facility will require five years from the change date, it will probably be necessary to move.

When comparing the space available to the space needed, keep in mind that when moving a large department, or several smaller departments, from an existing space to a new space, the remaining departments may not be able to use the vacated space efficiently. Consequently, it may be necessary to shift several departments in order to make the best use of the space vacated by one large department.

If it appears that there will be adequate space in the current location — after renovating and/or adding on — then a careful evaluation of the present building should be made in order to determine whether it will be more appropriate to stay or to relocate.

The following questions will help in making this decision:

- Location:
  Is the neighborhood good? Is it improving or declining? Is it convenient to where employees live?
- Accessibility:
  How accessible is the building to transportation, parking, shopping and other amenities, and to the disabled?
- Quality of the local work force:
  Is it skilled, unskilled, changing — for the better or the worse? What is the local work ethic?
- Transportation:
  Is there good public transportation available? What about ease of access by automobile, availability and cost of parking? What is the access by foot? How safe is the area surrounding the location? How accessible is the building for cartage deliveries or for over-the-road delivery off of interstate highways?

- Nearby amenities:
  Are there good restaurants? Fast-food and take-out restaurants? What about stores, dry cleaners, automotive service, banks or ATMs, etc?
- Proximity to business:
  What about the location in terms of proximity to present business? To potential business?
- The building itself:
  Is it well-managed? Does it present an appropriate image for the company? Is it well-maintained? How up-to-date are the building systems? If the building is not a new one, does it have reasonably sophisticated and adequate systems? Do the elevators operate on a programmed control system? Does the building have sufficient primary and secondary electrical power, and is it currently available to the floors the company is interested in? Does the heating and cooling system deliver adequate amounts of properly tempered and filtered air? Do the exterior windows have some form of solar control so that heating, cooling and lighting controls can be uniformly dealt with?
- How is security?
- What about fellow tenants? Do they add to the dignity of the company's public image—or detract from it?
- How difficult would it be to remodel the present facility to meet the company's needs? To answer this question, some preliminary block diagrams may have to be prepared of present space and of any proposed additional space. (Block Diagramming is explained in Chapter 5.) The preliminary block diagrams will show how extensive the reorganization will be. Knowing this, it is possible to begin assessing the extent of renovation necessary to facilitate this reorganization.

If the facility stands up well under evaluation, then the feasibility and cost of renovating can be discussed with the landlord. The main questions to cover are:

- How long could or should the new lease term be?
- What conditions would be attached to a renewed, extended, or expanded lease?
- How cooperative is the landlord prepared to be?
- What would the new lease costs be? (See section below for a discussion of calculating and comparing rentable areas.)
- What tenant allowances for renovation is the landlord prepared to make? (See section below for an explanation of tenant allowances.)
- Will the landlord provide a "fix-up" or renovation allowance for areas the company currently occupies, in the event a decision is made to either renovate or reorganize its existing space or expand it?
- More than likely, the landlord would prefer that the company remain as a tenant, rather than lose a tenant and go through the costs of releasing and finishing out a space for a replacement tenant. Therefore, the company is in a good position to ask for a "fix-up" allowance. This allowance should be sufficient to do a quality renovation. Make it clear to the landlord that the company may be willing to stay if the space can be renovated—at the landlord's primary expense.

When a general idea of conditions, terms, and costs has been obtained from the landlord, compare them to similar ones at competing buildings in the area—plus the cost of moving. Is the landlord trying to charge more rent than other comparable tenants in the building because he feels the company is "locked in," or is he making what appears to be a good and competitive offer?

In addition to determining the costs incurred in a new lease, it is necessary to estimate the cost of renovating the present facility, the cost of construction and furnishings in any additional space, and the cost of lost production time due to construction and/or moving.

If company space requirements projections indicate that the present location is not too small for its needs for the period of years covering a lease renewal, and if the location or quality of the present building satisfactorily represents the company's image and occupancy needs, the company should consider negotiating a lease renewal and staying in its current location. The cost savings will be substantial.

Renovation

If the decision is made to stay, the lease negotiation should be handled in nearly the same manner as defined in the following text for negotiating a new lease. This is the time to correct any lease conditions which were troubling during the prior lease term.

Follow the same procedure of progressing through a Letter of Intent to a Work Letter. In this instance, there is the complicating factor of trying to renovate spaces in progressive stages while relocating employees. It is possible that the landlord has some vacant space which could serve as "surge" or temporary space for some personnel while the renovation takes place. If not, employees will have to be asked for their patience with the renovation process. They may be surprisingly cooperative and helpful.

When considering renovation, do not be afraid to start with a clean sheet of paper. Ask the planner to address the renovation as if the space was totally vacant. Reorganize the space following the same planning, blocking and stacking process that would be used in a new facility. It may be surprising how much more efficient the existing space can be if no attempt is made to preserve old habits.

Expansion

Expansion may be just that—the enlargement of an existing space to accommodate a new division or group. Or, it may be in recognition of general overall growth. In either case, ask for a renovation or "fix-up" allowance from the company's landlord. If expanding to simply add a new group, and the existing facility is in good condition and operating effectively, the task is simplified. Do not try to redo a space that is functioning as programmed and designed just because the opportunity is there. On the other hand, do not hesitate to ask the question: Is this an appropriate time to renovate? Do a quiet survey of department heads and senior management. This may be the proper time.

Leasing Arrangements When Renovating or Adding Space

Remember that any existing renewal options were written into the original lease. Also written into the lease was a specific date by which the company must act if they choose to exercise an option. These dates should always be calendared in such a way that the company will be notified no less than twelve months before the expiration date of the renewal option. In this way, there will be enough time to evaluate the implications of exercising the option, and to make a well-considered decision.

If the assessment indicates that the company may be staying in its present facility, and it is discussing lease renewal terms, ask the building management if any other space will be coming available within the building during the time being considered as a renewal term. If some space will possibly be coming available, analyze it carefully. At the same time, ask the landlord to write into the company's lease a clause granting it first right of refusal for any space that becomes available that might suit its needs at a later date.

Remember, review the current status of the real estate market in the area. If there is a great deal of space available in other buildings, the landlord may be willing to upgrade the company's existing space at his expense in order to keep the company as a tenant.

If, on the other hand, little space is available and it is necessary to draw up a new lease with the present landlord, matters can become more difficult. Once local market conditions have been reviewed, lease negotiations can go forward. The following subsections review the information needed to analyze lease costs, terms, rentable area, usable area, tenant allowances, and building standard cost projections.

Locating Appropriate Sites If a Move Is Necessary

Surveying the Market

If a new site must be found, the process begins with a broad survey of local market conditions and space availability. Do not pick one particular locality or facility on the basis of "gut feeling." Such emotional decisions can prove extremely costly. There are at least three or four different types of information sources available. Contact people from all four. They are:

Developers

- Even if a particular developer does not seem to be working in areas that appear most desirable, he or she may be doing so behind the scenes—or may know others who are.

Owners

- Owners will often be willing to negotiate better rates than brokers. Also, some owners do not use brokers, preferring instead to handle all leasing themselves.

Commercial Brokers

- Most of them are familiar with the availability of space, even though they will not represent all of the spaces that are available; they should be able to identify brokers who do, or they may be able to co-broker a representation. Brokers are a good source of information and are usually worth their fees, which are normally paid by the landlord.

Professional Planner or Architect

- Many of these people may be working in new and existing buildings in the areas that interest the company. They may have a feel for what kinds of deals some of their other clients seem to be making. They probably will not know specifics, and if they do, they probably will not disclose such information. However, they will have a good feeling for the efficiencies of the building floorplate and the quality of the Work Letter and the building. They may have had experience with the contractors, and should be able to comment on their performance, cooperation and quality of work.

It may be necessary to contact five solid sources just to develop two or three good prospects.

New vs. Existing Buildings

Frequently, high-quality existing buildings are superior to new ones in terms of their location, price, established operating procedures and better-known reputations. Be sure to check with such buildings—space may be coming available in one which would be just right for the company.

Asbestos may be an issue in an existing building (generally only a problem in structures built between 1900 and 1973.) Ask the owner if the building has been surveyed for asbestos in the last five years, and ask to see a copy of the results of the survey.

"Indoor pollution" is a growing concern. Many building materials can emit gasses or particles that are irritating or even hazardous to building occupants, or when the building is undergoing a major renovation. The results of exposure to such materials can range from headaches and general weariness, to physical illness. On occasion this condition is so severe that portions of a building have been put out of normal use. Fortunately, there are environmental evaluation and treatment companies which can diagnose "sick building syndrome."

The effects of indoor air pollution seem to be strongest when the culprit materials are new; hence, they are not as often a problem in older buildings. There are also some techniques being tried today which may alleviate the problem in new buildings, such as "baking" the building before it's occupied, or only using building materials proven to be safe. Be certain to ask current tenants if they've had any problems that can be linked to indoor air pollution.

To help evaluate existing buildings, we have prepared an Existing Building Checklist shown in Figure 2.1. We have found that notes made at the time a space is inspected are infinitely superior to any "good-better-best" qualification system. Notes help you recall specific spaces and conditions in detail after you've inspected a large number of alternative sites. The form can be found under Master Forms. Simply make one copy of the form for each site you visit. Write your notes right on the form. That way you will have asked the same questions about each site.

You will find it helpful to photograph each building several times—inside and out—with an instant-developing camera. Attach the photographs to the appropriate Existing Building Checklist to assist you in recalling specific details later.

Urban vs. Suburban

To have a broad understanding of local market conditions, check Class A rates in both the downtown area and in the suburbs. Rates often vary widely. Parking costs should be substantially less in the suburbs, which may affect direct labor costs. However, lack of public transportation, greater travel distances and lack of proximity to business-related or peer businesses may indicate a nonsuburban choice.

Calculating and Comparing Rentable Areas

Landlords' definitions of rentable area (the square footage on which rent will be paid) may differ significantly between full floor and partial floor occupancy, from building to building, and by location, particularly from one region of the country to another.

Various associations of landlords and building management companies have developed standards for measuring floor space. However, no one of these is universally followed. When surveying the available space market, it is best to decide on one measurement system for all buildings. This is not as easy as it may sound, as many landlords will develop their own system in order to include in the "rentable" amount as much square footage as possible. Many questions must be asked to determine what is included in the landlords' quoted areas and rates. Establish benchmarks so that there will be some means of comparing the space under consideration.

Facility: _____   Location: _____

Department: _____   Date: _____

Prepared By: _____   Phone: _____

I. LOCATION

A. Accessibility

By public transportation? _____

_____

By private vehicle? _____

_____

By foot? _____

_____

For the disabled? _____

_____

B. Amenities Available

Nearby Restaurants	Number	Quantity	Proximity
Carry-out			
Fast food			
Sit-down			
Quality			
Nearby Shops and Shopping Services			
Barber and beauty			
Cleaning and repair, etc.			
Shopping centers			
Food markets			
Banking and ATM's			

Figure 2.1

Auto service _____

Drug stores _____

Parking _____

C. Neighborhood

Trend -- developing, over-developing, static, or declining? _____

_____

How compatible is it to the property? _____

_____

II. GENERAL APPEARANCE (AESTHETICS AND CONDITION)

A. Construction

Does the building have a quality appearance? _____

_____

Is it cheap looking? (Explain) _____

_____

B. Maintenance Standards

Repaired or patched? _____

_____

Up-to-date or planned? _____

_____

C. Exterior

Grounds

Landscaping -- neat and maintained? _____

_____

Drives

Repaired? _____

_____

Figure 2.1 (continued)

Orderly? _____

_____

Parking lots -- holes, cracks, striping, lighting? _____

Security? _____

_____

Walks -- even, orderly, security? _____

_____

Lighting -- working, adequate? _____

_____

Accommodations for the disabled? _____

_____

Buildings

Walls -- cracked, broken, clean? _____

_____

Windows -- cracked glass, operable, caulked, clean? _____

_____

Roofs -- blisters, patched, orderly? _____

_____

D. Interior Public Areas

Functional -- How well do they work? _____

_____

Cleanliness? _____

_____

Is the design contemporary, attractive, appropriate? _____

_____

Figure 2.1 *(continued)*

E. Toilet Rooms (Men's & Women's) (check both)

Cleanliness? _____

_____

How do they smell? _____

_____

Attractiveness? _____

_____

Equipment -- modern or out-of-date? _____

_____

Ventilation? _____

_____

Disabled accommodations? _____

_____

Security system? _____

_____

F. Hallways, Stairways, Corridors

Attractiveness? _____

_____

Lighting and signage? _____

_____

Cleanliness? _____

_____

How do they smell? _____

_____

How new do they look? _____

_____

Figure 2.1 (*continued*)

Disabled accommodations? _____

_____

Security? _____

_____

G. Elevators

   Attractiveness? _____

_____

   Lighting and signage? _____

_____

   Cleanliness? _____

_____

   How do they smell? _____

_____

   How new do they look? _____

_____

   Do they operate smoothly? _____

_____

   During a busy time, how long did you wait for an elevator on the ground floor? On the second floor? On the floor you're looking at? (Try this several times) _____

_____

   Accessibility for the disabled ? _____

_____

H. Building Service

   For storage _____

_____

Figure 2.1 (continued)

Rubbish removal? _____

_____

Service vehicles? _____

Access to building? _____

_____

Service vehicle parking? _____

_____

III. Tenanted Areas

A. Size

Sizes of spaces occupied by existing tenants (specify)

Square feet _____ to square feet _____

Average tenant space _____ square feet.

B. Tenant Improvements

How have other tenants improved and furnished their spaces? _____

_____

Dollars of investment: From $ _____/SF to $_____/SF (ask several).

_____

Quality of build-out and furniture. (Describe) _____

_____

C. Building Standards

Modern lay-in light fixtures? _____

_____

Modern lenses? _____

_____

Are they yellowed? _____

_____

Figure 2.1 *(continued)*

Suspended acoustic tile ceilings? Concealed or exposed suspension system? _____

_____

Permanent vs. demountable partitions?_____

_____

Walls -- paint or vinyl? _____

_____

Window treatment -- blinds or drapes? Condition? _____

_____

Floor coverings -- carpeting or vinyl tile? _____

_____

Comfortable room temperatures? Drafts? Stale air? Dampness?_____

_____

How do the rooms smell? _____

_____

What is the noise level?_____

D. Cleaning and Housekeeping

Finger marks on doors? _____

_____

Carpet dirty at edges? _____

_____

Windows streaked or cracked? _____

_____

Marks on walls? _____

_____

Figure 2.1 (*continued*)

Dirt around air diffusers on ceilings? _____

_____

IV . MECHANICAL EQUIPMENT
(Visit some mechanical and fan rooms and observe)

General organization of rooms? _____

_____

Clean and orderly? _____

_____

Lighting? _____

_____

Ask about service contracts -- full maintenance, in-house, preventative, or grease and oil? __

_____

Availability of plans and operating procedures? _____

_____

V. EMERGENCY AND LIFE SAFETY SYSTEMS

Manual alarm stations? _____

_____

Emergency lights?_____

_____

Lighted exit signs? _____

_____

Ionization, smoke or heat detectors? _____

_____

Public address system? _____

_____

Figure 2.1 *(continued)*

Fire extinguishers? _____

_____

Fire hose cabinets? _____

_____

Sprinklers -- code and I.S.O or Factory Mutual compliance? _____

_____

Disabled refuge areas, if not sprinklered? _____

_____

Figure 2.1 (*continued*)

Usually the term *rentable area* is that on which rent will be paid. This area may vary from floor to floor in the same building, depending upon whether the floor is a full floor occupancy or a multi-tenant occupancy. In addition, building common areas such as the main building lobby may be pro-rated into *rentable area* for all tenants.

Full Floor Occupancy

The rentable area of an office space which takes up an entire floor is normally calculated on a *full floor* basis, measuring from the glass line or the interior surface of the exterior building walls (depending on whether there is more glass surface or finished surface) to the finished surface of the core walls. Such obstacles as columns, convectors and window sills are counted as rentable areas even though they are not usable by the tenant. In addition, parts of the building's core are included in the landlord's calculation of rentable area, since the full floor tenant has sole use and benefit of them. These areas typically include toilet rooms, elevator lobbies, corridors, electrical and telephone closets, janitor's closets and any stairways used solely by the tenant. Under some circumstances, landlords will call this space "usable," even though portions of the space are not usable.

Partial Floor Occupancy

The rentable area of a partial floor is generally measured from the glass line or the interior surface of the exterior building walls (depending on whether there is more glass surface or finished surface) to the finished surface of the inside (tenant's side) of the walls of the public corridors, and from the centerlines of the partitions separating the space from neighboring tenants. These walls are usually called *demising walls*. As was the case with full floor occupancies, such obstacles as columns, convectors and window sills are counted as rentable areas even though they are not usable by the tenant. Again some landlords will term this space *usable*, preferring to measure all of the nonrentable space on the floor, such as toilet rooms, electrical closets, telephone closets, etc. as *rentable*. They will then set up a ratio relating the amount of space to be occupied by a tenant to the total *rentable*, and use the resulting ratio to increase the rented space.

In their calculation of rentable area, some landlords may wish to include a loss factor to recover the cost of such areas as major mechanical rooms, switchgear rooms, ground floor lobbies, atriums and special tenant amenities such as exercise facilities, conferencing centers and the like. This loss factor may be included in a tenant rentable area, even though these areas are located on totally different floors from the one the tenant would occupy. Another way for the landlord to recover these loss factors is to build them into the base rental.

Clearly, tenants can end up paying for more space than they will actually be able to use for the company's functions. This practice is, however, not without justification. An office building is a complex facility requiring many elements and systems that provide the tenant with various services. Such things as a telephone switch room located in a basement may be the point to which the telephone utility brings a tenant's telephone lines. From this point, the installing phone company may run cables to a telephone closet located in a core area on a multi-tenant floor, and then into the tenant's space. The tenant is using some portion of the basement telephone room, some portion of the telephone conduits which feed into the telephone closet on its floor, where the tenant makes another feed into its space.

Assignable Space

Due to the confusion of terminology in the marketplace, many individuals in the designing and leasing industry have come to apply the term assignable to the space which can be actually assigned to people, workstations, equipment and the like. It does not include intra-department or inter-department circulation.

Full Building Occupancy

If the rental of an entire building is under consideration, there are a number of additional factors to deal with:

- The building area may be calculated, on a gross square foot basis from outside wall to outside wall, with no deductions for anything—not for shafts, fire stairways, mechanical rooms, etc.
- On the other side of the coin, the tenant should be able to negotiate a more favorable rental and insurance rate with the landlord, especially liability insurance rates, on the basis of its sole occupancy.
- In full occupancy buildings, tenant allowances should be calculated on a *rentable* basis, from which the only deductions should be for space already finished by the landlord, such as toilet rooms, fire stairs, elevator shafts, mechanical rooms, etc.
- It is apparent that full building tenants will end up paying for more space than will actually be used for the company's direct operations.

In evaluating buildings that appear suitable, it is helpful to develop a matrix such as shown in Figure 2.2 which allows a comparison of rates over various lease terms.

Building Cost Comparisons			
PARTIAL FLOORS	**Building A $/SF**	**Building B $/SF**	**Building C $/SF**
5 years	18.00	17.80	18.50
10 years	18.00	17.70	18.00
15 years	17.80	17.70	17.65
20 years	17.30	17.30	17.65
FULL FLOORS	**Building A $/SF**	**Building B $/SF**	**Building C $/SF**
5 years	17.20	17.05	17.60
10 years	17.20	17.00	17.20
15 years	17.05	16.80	17.20
20 years	17.05	16.70	17.00

Figure 2.2

Rental rates are only part of the information needed to make a valid cost comparison. Assuming that all other factors, such as location and amenities, are equal, the ratio between rentable area and usable area may differ from building to building. To determine this ratio, it is necessary to calculate the usable area on each floor of each building under serious consideration. (If a landlord offers usable area figures, re-check them.) Multiply the result by the rental rate to obtain cost per usable square foot.

Example

The landlord says he will charge $18 per square foot for the space, and says that the rentable area for that space is 22,160 SF. Using the floor plans, it is possible to calculate that the space offers only 20,083 usable square feet.

$$22{,}160 \text{ SF} / 20{,}083 \text{ SF} = 1.1034$$
$$\$18.00 \times 1.1034 = \$19.86$$

Therefore, the effective cost for each usable square foot is $19.86.

Once the cost per usable square foot of each building under consideration is known, it may become evident that the building that appeared to be the most expensive (according to its rental rates per rentable square foot) actually turns out to be the least expensive and a good value (according to its rental rates per usable square foot).

Calculating usable square footage is important not only from a cost comparison point of view, but also from a space analysis point of view. Only when a particular building's ratio between rentable square footage and usable square footage is known, is it possible to determine how much space is needed for lease (in terms of rentable area), in order to obtain enough usable square footage to satisfy the company's space requirements. Evaluating these differences can be quite complicated. It may be prudent to hire an architect or space planning consultant for assistance in this regard.

Negotiating the Lease

A new lease can and should be negotiated, as should any lease renewal. The landlord's terms are not etched in stone and should be considered as his opening offer. Real estate values and rentals are affected by supply and demand probably more than almost any other commodity other than food and metals. So study the market. Negotiate.

Things to remember: A prospective tenant who is a large space user, has a prestige name, or has a very high credit rating is highly prized by a landlord, especially one trying to lease out a new building. A tenant willing to commit to a longer term lease has a special value to a landlord by adding credit value to his financing. The more committed, or leased-up the building becomes, the less negotiating power the tenant will normally have.

The shorter term lease, five years or less, may seem like a good deal to the tenant, and from a business plan standpoint may be a requisite. On the other hand, when the time comes for renewal, the landlord may be holding most of the good cards and may be able to negotiate from a position stronger than the tenant's. Think about those relocation costs: phone system, stationery, searching for and negotiating for new space, additional tenant improvement costs not paid for by the landlord, moving costs, lost personnel who are unwilling or unable to follow you, and finally, the lost business time suffered as a result of being "down" during the shift from one site to another. Refer to the section entitled, "Deciding If a Change Is Necessary," in Chapter 1, for more on this topic.

Note: *The steps described in "Estimating Square Footage Needed" (earlier in this chapter) must be completed before beginning any of the work described in this section.*

In addition to the five years from the change date growth factor, other possible growth factors should be considered. It might be in the company's best interest to execute a lease for 10, 15, or even 20 years. The company may wish to reserve space for future requirements by negotiating options that make additional space available for expansion at specified future dates. Alternately, the company may wish to lease excess space to ensure prospective growth needs. It may be possible to recover the cost of excess leased space (which will not be used for the next 5, 10, or 15 years) by subleasing it. (See "Negotiating Subleases On Expansion Space As a Landlord," later in this chapter.)

Preliminary Plan

During the search for new space, and especially during the negotiation stage, it is best to work with a preliminary floor plan. Prior to developing such a plan for any particular space, the estimate of square footage needs should be complete (see Chapter 1.).

A basic floor plan for a specific space should be developed while negotiating the lease. It should show reception areas, storage, general office areas and private office areas. How well present and future needs are satisfied by a particular space will have a large influence on the company's negotiating posture. Try to get the landlord to provide this space plan. It should cost about $.50 to $1.00/SF, and is well worth the effort.

Tenant Allowances

Another very complicated aspect of leasing space concerns tenant allowances—the physical finish (carpet, wall covering, lights, ceiling, doors, partitions, and the like) which a landlord agrees to provide the tenant as part of the base rent. The landlord will normally provide a document at some point during the early stages of negotiations which may be called by any one of a number of names, including:

- Improvements to Tenant's Suite
- Tenant Work Letter
- The Work Letter
- Building Standards
- Schedule B, The Work
- Outline Specifications
- Schedule of Rent Improvement Allowances
- Tenant Allowances
- Tenant Improvements (T.I.'s).

Whatever name is used, the matters covered by the document are of considerable importance, particularly when negotiating for large areas of space. Indeed, by negotiating favorable tenant allowances, it may be possible to save the company many thousands of dollars.

Analyzing tenant allowances is as important as analyzing rental rates. In some cases, a lease offer from one building which is lower than an offer from another (in terms of rent per usable square foot) may actually turn out to be more expensive when differing tenant allowances are figured in.

Shell Space

Before coming to a conclusion as to which Tenant Work Letter is best, it must first be established what the landlord is providing as the *base building shell*, the *shell space*, the *white box*. These terms attempt to define the base condition of the space, as completed by the landlord, before any so-called tenant improvement work is begun.

There are a vast variety of shell space conditions. Some will be relatively complete, some will barely meet the building codes for a temporary certificate of occupancy. A careful assessment of the shell space, as compared to the tenant allowances contained in the Work Letter, is a must.

The following are some shell space questions that should be asked. A careful record of their condition should be made.

- Are the interior columns finished (furred, enclosed in gypsum drywall, taped and sanded, ready for paint)?
- Are the exterior wall columns finished?
- Are the walls below the windows completed?
- Are there soffits above the windows to which a ceiling grid may be attached? If not, who installs them, and who pays?
- Are slot diffusers in the soffits for air handling, and how are they controlled for temperature control? Can they be divided in order to serve two offices, and what effect will such a division have on air noise? Who pays to alter the system to accommodate for partition locations to regulate the temperatures, and how is it to be accomplished?
- Are there window blinds and/or drapery rods in place? If so, who pays to re-cut or replace them if you wish to build a partition across where they are installed?
- How far into the space has the supply and return air ducting been run?
- Are fire dampers going to be needed, especially at the main return air duct? If so, who pays for them?
- How smooth and level is the floor, and who pays to "flash patch" it before your floor covering can be applied?
- How complete is the sprinkler (fire suppression) system? Who pays for its modification and completion? What is the building standard spacing and sprinkler head type?
- Where is the electrical distribution point for the space? What are the available building standard voltages and per square foot wattages?
- Who bears the cost of Life Safety electrical requirements, such as exiting lights, emergency lights, fire alarm stations, fire speakers and hook-up to the main building annunciator system?
- Are there wet columns for water and sanitary drains? If so, are they conveniently located to serve your needs?
- What provision is available for special air-conditioning requirements, such as higher voltages, condenser water, outside air supply and exhaust, and condensate drainage? How are they provided, if any, and at what cost?
- Are demising walls between the space and adjacent tenants built to the structural deck or only to an in-place ceiling grid? Who completes them to provide for security, air-conditioning and acoustical control, etc.?

It may be helpful to involve the company's space planner in defining what the shell space consists of and how it will relate to the landlord's proposed Tenant Work Letter. This is no place for skimping on the cost for good consultants, who regularly work with these issues and may be aware of what the current practice is in the community.

Tenant Work Letters

Customarily, a landlord will provide a prospective tenant with a document listing improvements he is willing to make to the space, if it is leased. The improvements, typically, should be described in terms of both quantity and quality. The format and method of presentation of such documents varies widely from landlord to landlord. We have reproduced three samples (Figures 2.3–2.5) to give a "feel" for these documents.

Sample Work Letter No. 3 is the most inclusive and a good briefing document. It covers issues that will exist, to some extent, in all situations where any tenant improvement work is to be performed. Study this example carefully.

SAMPLE WORK LETTER NO. 1
BUILDING STANDARDS

DEMISING WALLS: Demising walls will be from the floor to the above-floor structure and will be of 5/8" gypsum drywall with sound attenuation material.

INTERIOR PARTITIONS: Interior partitions will be steel studs with 1/2" gypsum drywall and will extend to the acoustical ceiling.

DOORS: The entrance door to the tenancy will be 3'0" x 7'0" x 1-3/4" solid core birch or steel. Interior doors are to be hollow core birch or steel.

DOOR OPENINGS: Door openings will be pressed steel, painted to match adjoining wall surfaces in a ratio of 1 door per 200 square feet.

HARDWARE: All hardware, including latch or lock sets to be Russwin and door closures to be provided on all doors to public corridors.

CEILINGS: Ceilings are of exposed-grid, suspension-type, using 5/8" fissured, acoustical, mineral tile and will be a nominal 8'9" above the floor.

FLOORING: The floor capacity is 80 psf live load (including partitions). All floors will be carpeted, in a color to be selected by tenant, with building standard carpeting, or provided with an allowance of $10.00 per yard for tenant's selected and installed carpeting.

LIGHTING: Lighting will be recessed 277 volt fluorescent 2 x 4 4-tube fixtures in a ratio of 1 per 70 square feet.

ELECTRIC SERVICE: Electric service capacity will be on an average ratio of 1 duplex wall outlet per 135 square feet. Standard non-wired telephone outlets will be provided at a ratio of 1 per 200 square feet. Switches are provided on a basis of 1 single pole switch per room, or per 250 square feet. Power poles may be provided when deemed necessary by the building. Special or primary service, such as 220 volt service, transformers, computer or other services which, in the building owner's opinion, will require separate service, shall be at the tenant's expense.

HEATING, VENTILATING, AND AIR-CONDITIONING: Heating and cooling are available year-round. The system consists of a hot or cold radiation and convection system, combined with forced air around the entire building perimeter. The interior, or core, portions of the building are supplied with an air mixture of fresh and building-return air, which is constantly filtered and temperature controlled. General temperature control is affected through a sensory system located outside the building which monitors both the outside air temperature and the sun's radiation to determine the proper balance of heating and cooling on sunny or shady sides of the building on a concurrent basis. Further individual control can be affected through thermostatic controls in the perimeter system.

PAINTING: All wall surfaces, doors, trim and convector covers shall be painted using not more than one color per room with semi-flat on walls and semi-gloss on doors, frames and convector covers with building standard colors.

WINDOW COVERING: One-inch venetian blinds are furnished for all exterior windows by the building.

CABINETS AND SHELVING: Library and storage shelving will be building standard options. If provided under the terms of the lease, they will be from building standard units at tenant's expense.

Figure 2.3

SAMPLE WORK LETTER NO. 2

_____

_____

_____

_____

re: Workletter covering lease

of _____

_____ floor in

_____

Gentlemen:

You (hereinafter called "Tenant") and we (hereinafter called "landlord") are executing simultaneously with this letter agreement, a written lease covering the space referred to above, as more particularly described in said lease (hereinafter called "the demised premises") in the building known as _____

_____ .

To induce Tenant to enter into said lease (which is hereby incorporated by reference to the extent that the provisions of this agreement may apply thereto) and in consideration of the mutual covenants hereinafter contained, landlord and Tenant mutually agree as follows:

1. a. Tenant, at Tenant's sole cost and expense, shall cause the following drawings and specifications to be prepared:
Complete, finished, detailed architectural drawings and specifications for Tenant's partition layout, reflected ceiling and other installations, for the work to be done by landlord under paragraph 2 hereof.

   b. Landlord, at landlord's sole cost and expense, shall cause the following to be prepared:
Complete mechanical plans and specifications, where necessary, for the installation of air conditioning system and ductwork, heating, electrical, plumbing and other mechanical plans, for the work to be done by landlord under Paragraph 2 hereof; cost of drawings and specifications for Non-Building Standard work shall be borne by Tenant.

   c. Any redrawings occasioned by Tenant and any requested changes in the plans shall be at Tenant's expense.

   d. All such plans and specifications are expressly subject to landlord's approval and shall comply with all applicable laws, rules and regulations. Landlord covenants it will not unreasonably withhold its approval.
Tenant covenants and agrees to cause said plans and specifications referred to in 1.(a) above to be delivered to landlord on or before _____ Landlord will cause said plans to be filed with the appropriate government agencies in such form (Building Notice, alteration or other form) as landlord may direct.

2. Landlord agrees, at its sole expense and without charge to Tenant, to do the following work in the demised premises:

   a. Supply and install landlord's Building Standard partitions, located where shown on said plans described in paragraph 1.(a).

   b. Supply and install, in the amount shown on Tenant's plans, landlord's Building Standard doors, frames and hardware, consisting of office door, latchset, and hinges for opening.

   c. Supply and install landlord's Building Standard acoustical tile ceiling throughout.

   d. Supply landlord's Building Standard electrical facilities sufficient for a connected load of 3 watts at 277/480 volts, three-phase, and 2 watts at 120/208 volts, three-phase, per square foot of each area served.

   e. Supply and install landlord's Building Standard recessed fluorescent lighting fixtures in the amount of one such fixture per each 80 square feet of usable area; Building Standard duplex receptacle wall outlets in the amount of one for each 150 square feet of usable area; telephone wall outlets in the amount of one for each 225 square feet of usable area; and Building Standard wall switches in the amount of one for each office and as required by Code in open areas. Electrical wiring within the demised premises for other than Building Standard items is not included.

Figure 2.4

f. Supply and install landlord's Building Standard zoned air conditioning system with reasonable duct work, and thermostats in an amount not to exceed one thermostat for each conditioning zone. Said system shall be designed to be capable of maintaining, within tolerances normal in first class office buildings and subject to density factors of not more than one person per 125 square feet nor more than 5 watts of electrical load per square foot of each area served, inside space conditions of 73 degrees F dry bulb and 50% relative humidity when outside conditions are 100 degrees dry bulb and 78 degrees wet bulb.

g. Supply and install landlord's Building Standard carpeting, glued-down, without padding, color to be selected by Tenant from landlord's standard colors; or 3/32nd inch vinyl tile floor, color to be selected by Tenant from landlord's standard colors.

3. If landlord further agrees to perform, at Tenant's request, and upon submission by Tenant of necessary plans and specifications, any additional or non-standard work over and above that specified in Paragraph 2 hereof, such work shall be performed by landlord or the builder at Tenant's sole expense, as a Tenant extra. Prior to commencing any such work requested by Tenant, landlord will submit to Tenant written estimates of the cost of any such work. If Tenant shall fail to approve any such estimates within (10) working days, the same shall be deemed disapproved in all respects by Tenant, and landlord shall not be authorized to proceed thereon. Tenant agrees to pay landlord, promptly upon being billed therefore, the cost of all such work, satisfactorily completed and in place, including fifteen percent (15%) of the cost for supervision. Tenant agrees that the same shall be collectable as additional rent pursuant to the lease and in default of any payment thereof. Landlord shall (in addition to all other remedies) have the same rights as in the event of default of a payment of rent.

4. It is agreed that notwithstanding the date provided in the lease for the commencement thereof, Tenant's obligation for the payment of rent under the lease shall not commence until landlord shall have substantially completed the work to be performed by landlord as hereinbefore set forth in Paragraphs 2 and 3 hereof; provided, however, that if landlord shall be delayed in substantially completing the work to be done by landlord as a result of:

a. Tenant's failure to furnish plans and specifications in accordance with paragraph 1.(a) hereof; or

b. Tenant's request for materials, finishes or installations other than landlord's standards or materials, finishes or installations undertaken to be furnished and installed by landlord in accordance with paragraph 3. hereof; or

c. Tenant's changes in said plans and specifications; or

d. The performance by a person, firm or corporation employed by Tenant and the completion of the said work by said person, firm or corporation, then the commencement of the term of said lease, and the payment of rent thereunder shall be accelerated by the number of days of such delay.

5. Tenant shall have reasonable entry to the demised premises for the purpose of installing Tenant's fixtures, equipment, installations or special decorations, all of which shall be performed in accordance with all applicable government regulations and Codes. Tenant shall submit plans and specifications to landlord for landlord's prior written approval before commencing Tenant's work. Tenant shall procure, or have landlord procure, all necessary permits before commencing Tenant's work. Tenant shall provide landlord with a schedule for Tenant's work to be performed, which schedule shall be approved by landlord. Worker's Compensation and public liability insurance and property damage insurance, all in amounts and with companies and on forms satisfactory to landlord, shall be provided and at all times maintained by Tenant's contractors engaged in the performance of Tenant's work, and before proceeding with Tenant's work, certificates of such insurance shall be furnished to landlord.

Such entry shall be deemed to be under all of the terms, covenants, provisions and conditions of the said lease except as to the covenants to pay rent. Landlord shall not be liable, in any way, for any injury, loss or damage which may occur to any of Tenant's decorations or installations so made prior to the commencement of the term of the lease, the same being solely at Tenant's risk.

If the foregoing correctly sets forth our understanding, kindly sign all copies of this letter agreement where indicated.

Very truly yours, _____   ACCEPTED: _____

Figure 2.4 (continued)

SAMPLE WORK LETTER NO. 3

Schedule "B"
Work

Article I

Except as may be otherwise specifically provided, landlord agrees to and will, at their sole cost and expense, perform, furnish, install and provide in the Demised Premises "Standard of the Building" items of work hereinafter set forth:

1. Partitioning
   Gypsum board (5/8") each side of 2-1/2" steel stud of which one layer shall extend to the underside of floor above; lined with sound blanket material; not less than STC 38; amount as required.
2. Doors and Frames
   Integral pressed steel, 16 gauge floor to ceiling (8'4") bucks; flush fireproof 3/4 hr. doors, hollow metal, full height; as required.
3. Closets
   Each closet 5'0" (nominal) in length, ceiling high, with flush bi-fold doors; two (2) wood hat shelves (paint grade) and one (1) metal chrome coat rod. One closet for up to each 3,000 square feet of net rentable area.
4. Hardware
   Latch sets (as manufactured by Schlage "D" series, or equal), 1-1/2 pair of olive knuckle ball bearing butts; wall door stops for each door. One lockset per floor or Tenant. Entrance doors to have concealed closer and lockset, Schlage "D" or equal; masterkeyed to Building system.
5. Painting
   All partitions painted three (3) coats, one primer and two (2) finish coats, alkyd. Exposed metal without factory finish to be painted one (1) enamel undercoat and one (1) semi-gloss enamel finish. Dark colors and additional colors per room will be an extra charge to Tenant.
5. Flooring
   Direct glued-down carpeting, without padding as manufactured by Bigelow, "Stati-Tuft" series, or equal; or vinyl composition tile floor covering 3/32" as manufactured by Armstrong, "Standard Excelon" series or equal; 4" rubber base on all walls.
6. Hung Ceilings
   Mechanically suspended acoustical ceilings, concealed spline, scored, 1'0" x 1'0", medium mineral fissured as manufactured by U.S. Gypsum, Accustone "F" or equal.
7. Lighting
   Four-tube fluorescent light fixtures (without lamps) 2'x4' as manufactured be Neo-Ray "Trim Troffer" or equal, one fixture per 80 square feet of net rentable area.
8. Window Covering
   One inch, natural color venetian blinds, in blind pockets, on all windows; no substitutions will be permitted.
9. Heating, Ventilating and Air Conditioning
   Year-round air conditioning system, combining the use of periphery units and interior duct air distribution system. The periphery air conditioning system shall service that portion of the Demised Premises being approximately 15' distant from the glass line of the building. For periphery rooms landlord will furnish one (1) thermostat for every two (2) peripheral bays. Each peripheral unit will be placed at each two (2) window modules so that each window will be served, eliminating the need for extra units due to subdivision of the premises. Interior duct distribution, including Standard of the Building ducts, grilles, etc. to accommodate Tenant's layout. The air conditioning system will maintain interior conditions of 80 degrees F dry bulb and 50% relative humidity when outside conditions are 95 degrees dry bulb and 75 degrees wet bulb, and will provide fresh air in a quantity not less than that required by Code, but not less than 0.35 cu. ft. per minute per square foot of floor area; all such conditions are provided that in any given room or area of the Demised Premises the occupancy does not exceed one (1) person for each 100 square feet, and total connected electrical load does not exceed 5 watts per sq. ft. for all purposes, including lighting and power. Duct distribution to accommodate Tenant's layout shall be designed only by an engineer in landlord's employ.

Figure 2.5

10. Electrical Outlets
    Duplex receptacles in Standard of the Building partitions, peripheral enclosures at prefabricated knockouts; or on floor over activate electrical cell; one (1) per 100 square feet of net rentable area.
11. Ceiling fixtures will be serviced by switches as required by Code, but not less than one (1) switch per room.
12. Telephone
    landlord will supply not more than one (1) point of telephone distribution for each 150 square feet of net rentable area, through peripheral knockouts, floor heads, etc. and shall make available empty duct, raceway, and/or conduit up to one (1) inch as needed. All wiring shall be done by others. No additional facilities conduit, raceways or wiring will be provided for intercom systems, special communications systems, or computer systems, whether public or private, except as provided in Item 14.
13. Electric and Telephone Capacity
    Capacity of the distribution system shall be sufficient for Items 11 and 13 for a continuous demand load of two (2) watts per square foot drawn at any power factor down to 80%; sub-floor cellular distribution contains three (3) 3" cells at least 6'0" on center.

Article II
Tenant shall on or before the _____ day of _____ , 199_ (herein called the "Submission Date") submit to landlord final and complete dimensioned and detailed plans and drawings of partition layouts, including openings, ceiling and lighting layouts, colors, and any and all other information as may be necessary to complete the Demised Premises, all at Tenant's cost and expense; it being specifically understood that the landlord does not provide architectural or layout services.

Article III
The initial air conditioning duct distribution will be designed at landlord's expense. In the event that any engineering services are required for any other purposes or as a result of changes in any of Tenant's layout or Tenant's requirements for occupancy, Tenant shall reimburse landlord for the cost of such engineering services. Such engineering services shall be performed by landlord's engineer only.

Article IV
All plans, drawings and specifications with respect to the Demised Premises and required to be submitted by Tenant to landlord, shall comply with and conform with landlord's Building plans filed with the Department of Buildings, and shall comply with all Regulations and Codes and/or other requirements of any governmental department or agency having jurisdiction over the construction of the Building and/or the Demised Premises. Any changes required by any governmental department or agency affecting the construction of the Building and/or the Demised Premises shall be complied with by the landlord in completing said Building and/or Demised Premises and shall not be deemed to be a violation of the Tenant's plan or any provisions of this Schedule, and shall be accepted by Tenant.

Article V
Notwithstanding anything to the contrary contained herein, if Tenant shall fail to comply with the Submission Date as set forth in this Schedule, landlord may proceed with all work required to be done by it in accordance with the plans and drawings last submitted by either party to the other and/or in accordance with the Standard of the Building and such completion shall be deemed to be substantial completion of the Premises in accordance with the provisions of the Lease.

If within ninety (90) days after such Submission Date, no plans are submitted by Tenant, then landlord may complete the Demised Premises only in accordance with the minimum legal requirements necessary for occupancy as "open space," and completion in such manner shall be deemed full compliance with this Schedule, and landlord shall be released of all further obligations under this Schedule. Such failure on Tenant's part to submit Tenant's plans within 90 days after the Submission date shall be deemed a default under the Lease and landlord shall have all rights and remedies as provided therein.

Article VI
Tenant does herewith further agree that in the event substantial completion of the Demised Premises or the substantial performance of the work to be performed by the landlord pursuant to this Schedule and/or any further agreements pertaining thereto entered into between landlord and Tenant is delayed by Tenant's failure to comply with its obligations under this Schedule or by Tenant's changes, Tenant (in

Figure 2.5 (continued)

addition to paying for the costs and damages landlord may sustain by reason of such changes and/or by reason of delay in submitting such layouts) agrees that the term of this Lease shall commence on the date that this Lease would have commenced had not the completion of the work been so delayed by Tenant.

Article VII

Tenant may specify a substitution of any item of Standard of the Building work, and of any item of work set forth in Article I, excepting those items for which substitution is specifically prohibited, provided that (a) such substitution shall be of a like nature and of equal or greater cost and quality than that for which it is substituted; and (b) landlord shall install such substitute. Tenant shall pay the cost of such substitution, but shall be entitled to credit for that for which such substitution was made, based on landlord's cost thereof, and further provided that in no event shall Tenant be entitled to apply such credit except against the cost of the item thus substituted, and in no event shall Tenant be entitled to be paid in cash for any credit, nor shall Tenant be entitled to credit for items omitted or not installed.

Article VIII

Except as is otherwise herein permitted or required, any approvals or disapprovals required to be given by either party hereto to the other shall be given within ten (10) calendar days (holidays excluded) after written request thereof. Submission of shop drawings, plans, layouts, etc. and requests for authorization to proceed with the work at Tenant's expense, but not disapproved or corrected within ten (10) calendar days, shall be deemed approved.

Article IX

In the event that Tenant shall desire to perform and make any installations (hereunder called "Tenant's Installations") which are not to be performed by landlord for Tenant, landlord agrees to afford Tenant access to the Demised Premises prior to the completion and possession date for the purpose of making inspections, taking measurements and making Tenant's Installations (all of which are to be paid for by Tenant), provided that such Installations will not require any structural change, and further provided that the construction of the Building and the Demised Premises and all installations required to be made by the landlord therein shall have reached a point which in landlord's sole judgment, expressed in good faith, will not delay or hamper landlord in the completion of the Building and/or the Demised Premises. Prior to the commencement of Tenant's Installations, Tenant shall submit to landlord complete detailed plans and specifications of the work for landlord's prior written approval. Any entry by Tenant in or on the Demised Premises shall be at Tenant's sole risk, and upon request of landlord, Tenant shall pay for and deliver to landlord policies and certificates of insurance in amounts and with such companies as shall be reasonably satisfactory to landlord, such as, but not limited to Public Liability, Property Damage and Worker's Compensation to protect landlord and Tenant during the period of making such Tenant's Installations. landlord shall be named as an insured party in such policies or certificates of insurance and the same shall be continued in effect during the period of the performance of Tenant's Installations. All Tenant's Installations shall be in accordance with the rules and regulations of any governmental department, bureau or agency having jurisdiction thereover, and shall not conflict with, or be in violation of, or cause any violation of landlord's basic plans and/or the construction of the Building and all of Tenant's Installations shall be completed free and clear of all liens and encumbrances. All permits which may be required by Tenant for Tenant's Installations shall be procured and paid for by Tenant only after having obtained landlord's written approval for such work, or, if landlord shall deem advisable, landlord may procure such permit, and Tenant shall pay for the same. No plans and/or specifications required to be filed by Tenant pursuant to any work contemplated to be performed by it within the Demised Premises shall be filed or submitted to any governmental authority having jurisdiction thereover without first having obtained landlord's approval to the same. Notwithstanding anything to the contrary contained hereinabove, landlord reserves the right to deny Tenant or its contractor access to the Demised Premises and/or to request Tenant to withdraw therefrom and cease all work being performed by it or on its behalf by any person, firm or corporation other than landlord, if landlord shall, in its sole judgment, to be exercised in good faith, determine that the commencement and/or the continuance of Tenant's work shall interfere with, hamper or prevent landlord from proceeding with the completion of the Building and the Demised Premises at the earliest possible date. Tenant agrees that should the Tenant enter upon the Demised Premises for the purpose of performing any work, the labor employed by Tenant or anyone performing such work for or on behalf of Tenant shall always be harmonious and compatible with the labor employed by landlord. Should such labor be

Figure 2.5 (continued)

incompatible with landlord's labor as shall be determined by the sole judgment of the landlord, to be exercised in good faith, landlord may require Tenant to withdraw from such premises until the completion of the Building or Demised Premises by the landlord. In the event Tenant or Tenant's contractor shall enter upon the Demised Premises or any other part of the Building, as may be above permitted by landlord, Tenant agrees to indemnify and save landlord and the Premises free and harmless, from and against any and all claims arising from or out of any entry thereon or performance of said work and from and against any and all claims arising from or to arise from any act or negligent act of Tenant, its contractors, agents, servants or employees, or from any failure to act, or for any other reason whatsoever arising out of said entry or such work.

Figure 2.5 (continued)

In cases where the company's space requirements exceed those of the average 5,000-20,000 SF user, the company is in a position to negotiate a more generous tenant allowance than the one that would normally be provided by the landlord. In addition, if space standards dictate an open or a modified open plan design, the need for such items as partitions, air handling diffusers and return air grilles will be greatly reduced. Naturally, the company will want to recover or shift those cost savings to another category, for its own benefit, rather than letting them remain the landlord's profits.

Negotiating these matters requires an extensive knowledge of current building costs and market conditions. The assistance of an architect or space planner can be important in negotiating the most favorable terms. Landlords are generally very cagey about tenant improvements, and they will negotiate very hard over these allowances. Sometimes they do not have an abundance of funds available. In other cases, they will almost "buy" a tenant with their generous tenant improvement allowances. It is a good idea to get all the help available in dealing with these matters.

Before final negotiations can begin, detailed information must be obtained about the basic tenant allowances from the landlord of each building under serious consideration. Since work letter documents vary so widely, the easiest approach may be to send a letter to the landlord of each building being evaluated. The letter could request him to restructure the tenant allowance information into a specified format. Responses organized in comparable terms should help in comparing competing offers. The letter in Figure 2.6 is an example.

EXAMPLE LETTER TO LANDLORDS REQUESTING TENANT ALLOWANCE INFORMATION

To: Name and address of landlord

Dear _____

We have reviewed your tenant installation specifications. It is our opinion that the interests of all parties involved could be best served by restructuring your (tenant Work Letter, tenant Allowances, etc.) to more appropriately reflect our anticipated space planning direction.

Present concepts indicate that a large percentage of the area we will require will be developed for open office planning. We feel strongly that this planning direction will greatly simplify the distribution of the basic systems required to service the open plan.

Therefore, we recommend that the basic shell development be revised to include the following:

1. Heating, ventilating and air-conditioning to be complete per tenant's requirements, with no additional charges.
2. Electrical service to be complete at voltages required, including all panels and distribution, with no additional charges.
3. Life safety systems to be complete with all exit signs, sprinklers (if required), annunciators, alarms, communications devices and emergency lighting requirements with no additional charges.
4. Window coverings to be complete with installation, with no additional charges.
5. Acoustical ceiling panels and suspension system to be complete per tenant's requirements, with no additional charges.

Further we recommend that the tenant Improvements package include the following items, complete without additional charges.

1. Walls, partitions, wall base and finishes.
2. Doors, frames, hardware and finishes.
3. Floor coverings (carpet as standard and vinyl composition tile as appropriate).
4. Light fixtures and switches.
5. Electrical service and telephone outlets from base building system to workstations, as required.

We are also concerned about core areas for floors which we may fully occupy and recommend that all core area materials and finishes for toilet rooms, elevator lobbies, etc. be subject to approval by us.

If your approach to the improvements is other than turnkey, we would appreciate a listing of the quantities of each of the above items which will be allowed, the precise specification for each, as well as the unit prices for additional quantities and credit amounts for unused quantities.

We look forward to receiving your proposal and would be happy to meet with you to discuss these recommendations.

Sincerely yours,

_____

Figure 2.6

Ideally, the responses you receive to your letter should look something like the samples in Figure 2.7. (The dollar amounts are for sample purposes only.) A form for this comparison may be seen as Figure 2.8 and found in the Master Forms Section.

Note: This example is based on prices prevailing in the Midwest area in Winter, 1990. Regional variations and continuing inflation make it highly likely that the costs shown here will not prevail for another, later job. Your space planner, architect, interiors contractor, or broker should be able to help you ascertain the current rates for your area.

TENANT UNIT COSTS

Building Standard Item Description	Quantity Provided per Rentable Square Foot	Unit Cost by Trade	Unit Cost per SF
Drywall Partitions 2-1/2" mtl. stud; 5/8" gyp. bd, each side, taped and sanded; to underside of 9'0" ceiling; with vinyl base; no paint.	1 LF per 15 SF	$25/LF with base no paint	$1.67
Entrance Doors at Elevator Lobby 5'0" x 8'4" x 1-3/4", oak, fire-rated, stain finished; frame; lock set, bolt butts, strike & closer; installed.	1 ea. per 7,098 SF	$2,600/EA	.37
Interior Doors 3'0" x 8'4" x 1-3/4", oak, solid core, stain finished; 2" steel frame; latch set, butts, and strike; installed.	1 ea. per 330 SF	$745/EA	1.05
Acoustical Ceilings 2' x 4' mineral board; incl. suspension; installed.	100%	$1.25/SF	1.25
Floor Covering carpeting glued down, no padding; base in partition allowance; installed.	100%	$15/SQ YD	1.67
Light Fixtures 2' x 4' lay-in, acrylic lenses, return air slots, with lamps; installed.	1 ea. per 75 SF	$110/EA	1.47
Duplex Electrical Outlets in Wall, multi-circuited to electrical closet; installed.	1 ea. per 125 SF	$80/EA	.64
Duplex Electrical Outlets in Floor, multi-circuited to electrical closet; installed.	1 ea. per 125 SF	$325/EA	2.60
Telephone Outlets in Wall to telephone closet; conduit only installed.	1 ea. per 150 SF	$30/EA	.20
Telephone Outlets in Floor to telephone closet; conduit only installed.	1 ea. per 150 SF	$275/EA	1.83
Switches toggle, single pole, 15 amp; installed.	1 ea. per 150 SF	$80/EA	.53
Paint 2 coats; oil base; wall surface including trim. (9 ft. high walls)	100%	$ .35/SF	.21
Wall Covering "II" quality; installed. (9 ft. high walls); option to paint	100% wall surface	$1.35/SF	[.81] [extra]
Total Value of tenant Allowance			$13.49

Figure 2.7

Once both the tenant allowances and the building standard cost projections are known, it may be easiest to compare costs through the use of a matrix similar to the one created to compare rents. Although not necessarily representative of current or local market conditions, the matrix might look something like that shown in Figure 2.7. A form for this matrix is shown in Figure 2.8, and will be found in the Master Forms Section.

When a lease is finally signed, the results of all the negotiations will be incorporated into it, either as a section within the lease itself or, more commonly, in a Work Letter attached to the lease as an addendum or exhibit.

Negotiating the Work Letter

Before examining the Work Letters, it is important to consider a number of questions which can and will have major economic impacts on tenant improvements. Have in mind the probability that the tenant Improvement Allowance will not cover all of the wanted or needed improvements. The tenant Work Letter has been constructed to protect the landlord and to limit his financial exposure. The risk of "extras" or "cost overruns" will be the tenant's, unless the company carefully negotiates the terms of the Work Letter. The landlord will not volunteer anything; it must all be negotiated.

The Letter of Intent

The best time to negotiate for the Work Letter terms is during negotiations of the Letter of Intent terms, such as the description of the demised space, rentable area, base rent, escalations or pass-through terms, base tax years, the lease commencement date and the lease term, options, special concessions, and other basic business terms. Both the Letter of Intent and the Work Letter should be included as exhibits to the Lease. The landlord will never again be more willing to negotiate these terms. The company's negotiating team should include members who have current and extensive backgrounds in leases, comparative properties, tenant improvements, construction and construction costs, planning and design, and a special awareness of the amounts and types of improvements the company as a tenant will most likely want or need. Do not be concerned about having a large team. The landlord has already spent months, perhaps years, preparing for this type of negotiation.

The Work Letter

Once the "shell space" has been defined, it is helpful to secure a copy of the landlord's "Building Rules and Regulations" which may contain a variety of conditions that will affect the company's design and use of the premises (such issues as hours of operation – after-hours heating and cooling, elevator services, cleaning, ingress, etc.).

The company as prospective tenant should also have examined the landlord's proposed Work Letter to determine where the "holes" are between what the letter actually says and what the landlord is saying he will do.

Work patiently and systematically from a prepared list of concerns and questions. Document each point in terms of both agreement, or lack thereof, and in economic values. Do not make hasty concessions in order to win some other point.

Plans and Construction Documents

If the landlord is to pay for "plans or construction drawings," or for engineering plans, will this include the company's space plans and designs? If so, for how many revisions and changes? It is likely that the intent of the Work Letter is for one plan. Without asking and negotiating, all revisions may be at the company's cost. It is best to negotiate a reasonable number of revisions, not only to your plans, but also for the engineering plans, especially mechanical, electrical and sprinkler. Will the "Plan Allowance" cover

Building Standards and Tenant Allowance Comparisons

Facility: **Branch** Location: **Fairview, IL**

Prepared By: **David Johns** Phone: **555-8989** Date: **11/22/93**

SF = square feet

Item	Building A			Building B			Building C		
	Tenant Allowance	Unit Cost	Value /SF	Tenant Allowance	Unit Cost	Value /SF	Tenant Allowance	Unit Cost	Value /SF
Partitions	14LF/100SF	$22/LF	$1.57	20LF/100SF	$19/LF	$.95	15LF/100SF	$21/LF	$1.90
Doors	1/200 SF	$380/LF	1.80	1/250 SF	$400 Each	1.60	1/200 SF	$300 Each	1.50
Flooring	$12.00/Yard	$1.33/SF	1.33	$9/Yard	$1/SF	1.00	$13/Yard	$1.44/SF	1.44
Lighting	1/70 SF	$75 Each	1.07	1/65 SF	$70 Each	1.08	1/75 SF	$80 Each	1.06
Electrical Wall Outlets	1/135 SF	$35 Each	.26	1/150 SF	$35 Each	.23	1/140 SF	$40 Each	.29
Electrical Floor Outlets	0	$275 Each	0	0	$280 Each	0	0	$285 Each	0
Telephone Outlets-Wall	1/200 SF	$25 Each	.13	1/250 SF	$30 Each	.12	1/200 SF	$30 Each	.15
Telephone Outlets-Floor	0	$150 Each	0	0	$135 Each	0	0	$155 Each	0
Switches	1/500 SF	$30 Each	.06	1/500 SF	$45 Each	.09	1/500 SF	$50 Each	.83
Paint	2 Coats	1 Coat $1.80/LF	.26	2 Coats	1 Coat $1.80/LF	.23	2 Coats	1 Coat $1.80/LF	.26
Vinyl	0	B Quality $3.60/LF	0	0	C Quality $3.00/LF	0	1LF/1OLF Partition	B Quality $3.50/LF	.05
Totals		$6.48			$8.11			$9.38	

Figure 2.8

improvements in excess of the "Building Standard Allowance"? Who will pay for the cost of designing special details such as special millwork, etc.? Who will pay for plans for such areas as computer rooms, special dining areas and for general design which is in excess of the standard allowances? Will the landlord charge an administration fee for assisting in the preparation and review of the "plans"?

Unit Costs

If the landlord is going to build the space with his contractors, he will be doing so with money the company, as tenant, has negotiated as its "tenant Improvement Allowance," in other words, with the company's money. It is necessary to work out, unit by unit, the value of the original tenant Allowance, and to be sure that you have the right to allocate those funds in any manner you wish. For example, do not let the landlord tell you that you must use all of his allowed quantity of building standard partitions, and that if you do not, you will risk losing the allowance. Instead, establish the unit value of the partitions with the landlord, and then determine if he will provide the company with credit for lineal feet of partition not used. Will he charge the exact same cost for additional lineal feet of partitions that the company may need to build?

If the landlord is willing to agree to a specific sum of money for all tenant improvements, the question then becomes: What will he charge for building standard units and for units in excess of the allowance that the company may wish to have built and to pay for? Company representatives should remember that this is the company's money, and should be sure to establish some working arrangement for the payment of Work as it progresses. Work may include not only the landlord's funds, but also funds for "over standard" work. Arrange for a mutually agreed upon representative to "sign off" on and to approve all invoices presented for payment; and especially to approve all change orders and extras which the company representative may have knowingly or unwittingly approved. An iron-clad procedure should also be established for extras and change orders to be charged to the company. It will be presumed that charges and extras are to be paid for by the company. Be sure that they are real extras and that all Work that should have been performed, and may not have been due to the changes, has been credited.

If the Landlord Performs the Work

For new construction, the landlord will usually do the Work, using the allowance and the tenant's own funds. The landlord will usually ask that the tenant provide its funds before the Work commences. The tenant will want to provide its funds at the end of the job. The best negotiating philosophy to follow is: pay for only what you have gotten, not something you may get. The solution can follow a variety of approaches. The landlord and the tenant can each deposit their funds into an interest-bearing escrow account. The landlord's contribution will likely be well-established and based on the amount determined in the Letter of Intent, or Work Letter. The tenant's contribution will be based on a best estimate. Not until the job is complete will the tenant's final costs be known.

Should the landlord be chosen to perform the tenant improvement work, using the tenant's allowance and the tenant's own funds, he has become the tenant's Project Manager. The tenant should deal with the landlord as such. The tenant should negotiate what is tantamount to a Project Management Agreement with the landlord. This agreement can be a portion of the Letter of Intent, a separate document, or a part of the Work Letter. It should define the duties, in the form of a "scope of work," the liabilities and responsibilities of the landlord/project manager. When negotiating this portion of the Work

Letter, there are some very large open windows out of which the tenant's allowance can fly if appropriate rules are not established in the Letter of Intent or Work Letter.

Landlord's Administration Fees

Administration fees or charges are a common cost charged by landlords for handling the relations with the planners, engineers, contractors, suppliers and regulatory agencies. This fee will usually range from 10% to 25% of some portion of the work. Some landlords charge an administration fee on all Work, plan review, Building Standard Work, and Over Building Standard Work. Other landlords will charge administration fees on only that portion of the cost of Work which exceeds the Building Standard Allowance. Try to eliminate it entirely. The landlord has already budgeted a Developer's Overhead and Fee for itself. If you cannot eliminate it altogether, try to limit the amount of administration fees charged for Work in excess of the Building Standard Work. Again, try not to pay an administration fee on the original tenant allowance funds. If you must pay a fee for Work above the building standard, one way to do this is to use a fixed amount for the initial over-standard amount, and a lesser fixed amount per change order. Try for an agreement that a change order may contain more than one item and will carry only the one administration charge. A charge of $250 to $500 should cover the landlord's administrative costs.

Contractor's General Conditions

Contractor's General Conditions represent another charge that is usually a percentage markup applied by the landlord's contractor to additional cost of Work. It covers the contractor's project management, supervision, site costs, etc. The usual contractor's percentage for general conditions is in the 5% to 15% range. The tenant should attempt to eliminate this markup entirely, unless the construction schedule is substantially extended as a result of the change orders. If there is no extension of schedule, the contractor is not suffering additional General Conditions costs and therefore is likely to be not entitled to additional compensation. General Conditions is an item that should be carefully examined at the outset to be sure that it does not contain costs which are already being covered in the normal "Costs of Work" from which the Unit Costs have been derived. Be sure that the landlord agrees that a General Conditions fee will not be increased unless the scheduled construction time is increased as a result of tenant delay.

Contractor's Overhead and Profit

Contractor's markup for Overhead and Profit is another area of exposure. Find out what this markup will be. Have the landlord agree that the markup will not exceed an agreed upon percentage. In this regard, try to find out where the contractor is carrying payroll overhead, or burden. When calculating Unit Costs, the contractor will most likely cover his labor burden in the Labor portion of the unit cost. He may then try to cover it again in the Overhead portion of the "Overhead and Profit" markup. The normal overhead markup is about 10%. The normal profit percentage is another 5% – on top of the 10% overhead markup. Both Overhead and Profit are typically in addition to General Conditions. The landlord's Administration fee is applied on top of all of the contractors' fees. The compounded total of these markups can be in the range of 25% to in excess of 45%.

Scheduling

Establishing and negotiating the rent commencement date will be one of the most critical, complex, and difficult portions of the lease. It is unlikely that a firm rent commencement date can be determined when negotiating the Letter of Intent. However, the tenant should realize that the landlord wants the tenant to begin paying rent as quickly as possible. The landlord will do anything it can to induce the tenant to commit to a fixed rent commencement date. A part of this negotiation will deal with two terms which are at once similar and very different: the *Certificate of Occupancy*, (or *Temporary Certificate of Occupancy*), and *Substantial Completion*.

Certificate of Occupancy

A Certificate of Occupancy is normally issued by the city through its Building Department. It merely authorizes someone to occupy and use the space. It is a recognition that the Work, as defined in the plans approved by the Building Department, has been completed and that the space meets all Building, Fire, Life Safety and other appropriate codes. Some jurisdictions will only issue what is called a *Temporary* or *Provisional Certificate of Occupancy* until the building becomes close to 100% completed. This is quite common when a building is not totally completed and occupied, in which event a Building Department may issue a temporary C. of O. for particular spaces within the building pending total or near total completion, at which time the Building Department will issue a final Certificate of Occupancy. The Certificate of Occupancy does not deal directly with the question of Substantial Completion.

Substantial Completion

Substantial Completion is a contractual term and generally means that the Work has been completed to a point that the occupant may use the space in the manner for which it was intended. It is possible to have a C. of O. without having achieved Substantial Completion. Likewise, it is possible to have Substantial Completion and not have a C. of O.

Dealing with both the Substantial Completion and the Certificate of Occupancy language will require the involvement of both the company's attorney and the rest of the project team. Once a tenant has committed to a Rent Commencement date, there are only a few events, such a force majeure, which can change it. The project team members should recognize that space planning, preparation of construction documents, approvals, bidding, negotiating, building inspections, deliveries, construction, special details, and cleanup all take more time than expected. Set a realistic target for the Rent Commencement Date.

If the Tenant Does the Work

Sometimes the landlord will permit the tenant to do its own work, with the draws to be paid out of a landlord/tenant escrow account. Or, the tenant may negotiate to receive the tenant allowance in the form of cash and do its own work. This latter procedure of receiving the "allowance" in the form of cash should only be carried out with the approval of the company comptroller or tax advisor, as it may result in the receipt of ordinary income without a commensurate offset in expenses. Placing the funds in an escrow account, out of which construction and other cost draws can be made, may be a better procedure; again, using an agreed upon representative to approve all invoices. This becomes of special value if the cost of the Work will exceed the allowance and the tenant must pay for the excess costs.

Lease Terms   Real estate leases cover a number of areas in which the final negotiated terms can benefit the tenant – or be very costly. The principal goal in lease negotiations is to shift to the other party the risk of having to pay additional monies in connection with the lease as a result of the occurrence of unexpected events. Because that is the goal, almost all so-called *legal* issues are really *business* issues; that is, they involve an attempt to limit the responsibility of one party or the other to pay for unforeseen costs. This in no way suggests that the company should not involve its attorney in developing the terms. The discussions which follow are written in business language; the actual lease will probably need to be written in legal terms. Therefore, the company's attorney should most assuredly prepare or approve the lease language and advise the company of local law and customs.

The following list indicates the contract terms that are likely to be most favorable to the company as tenant.

Parties

With whom is the company dealing? Consider obtaining a preliminary title report to make certain that the person who purports to be the landlord really owns the property. Generally, the name of the landlord in the lease will be the same as the name of the owner of the property. If there is more than one owner of the property, all owners should have to sign the lease as a landlord. Ask if the landlord with whom the company is dealing owns or leases the building. If the landlord has a leasehold interest in the property through a ground lease or master lease, review the ground or master lease to make certain that the lease the company is planning to sign does not conflict with the terms of the ground lease or master lease. If the landlord is a corporation, obtain a corporate resolution of the landlord that states that the lease is authorized and that the persons who are signing the lease on behalf of the corporation have the proper authority to do so.

Premises

Make certain that the premises are clearly defined in the lease. The number of rentable square feet the company is being charged for should be stated, and the term *rentable square feet* should be clearly defined (does it include non-usable space such as bathrooms, elevator lobbies, utility rooms, and the like?). The definition of premises should also include all other rights the tenant will have in connection with the property, such as parking rights, normal and after-hours operating costs, and rights to use of the common area. If parking spaces are granted, attempt to have those spaces specified as exclusive to the company and clearly designated on a map or drawing attached to the lease.

Improvements

If the landlord is going to construct improvements for the company, make sure that the plans and specifications for those improvements are spelled out in detail so that there will not be a dispute later of what the landlord was supposed to provide. This part of the lease will become an exhibit to the lease and will be referred to as the "Work Letter." The terms of the Work Letter must be completely defined and reduced to specific quantities, dollar allowances, and qualities before the lease can be executed. Under no circumstances can a lease be executed on the basis that these matters will be worked out after the lease is signed. (See previous sections in this chapter for more detailed information on assessing tenant allowances and the Work Letter.)

Term

Short-term leases (up to ten years) provide the tenant with the opportunity to move if the area surrounding the building should decline in quality or if a major change in space requirements should make a move desirable. However, short-term leases also permit the landlord to raise the base rent at the time of renewal. The best solution is to have a short-term lease with renewal options at specified rental rates. Attempt to obtain as many options to extend the lease as possible. However, have in mind that options are beneficial to tenants only. The landlord will not be anxious to extend them.

The lease should specify that its term does not commence until the landlord has completed all of the tenant improvements that he has agreed to provide, if any. In addition, attempt to negotiate a clause that reserves the right to cancel the lease if the tenant improvements are not completed by a firm date. Often the landlord will agree to such a provision, but will insist that the fixed date be extended for delays beyond the landlord's control (a so-called "force majeure" clause). If as tenant, the company agrees to such a provision, it should, in turn, insist on an outside date by which the tenant improvements have to be completed, or it can cancel the lease.

Rent

Always attempt to obtain a basic rent clause that contains no escalation provision. If, however, the landlord insists it contain an escalation provision tied to the cost of living index (a so-called "CPI" index), or some other index, try to insist that a maximum increase or "cap" be inserted in the lease; otherwise, the company will have no way of predicting rental obligation in the future.

If the lease contains a provision requiring the tenant to pay a prorated share of the building's operating expenses (a so-called "pass-through" clause), insist that the company be required to pay only actual increases in operating expenses. Make certain that the landlord maintains an adequate bookkeeping system and that the company has the right to inspect the accounting records. Always try to resist a clause which ties increases in operating expenses to a cost of living index or to any other index.

Tax Clause

If there is a clause in the lease requiring the tenant to pay either prorated real estate taxes or any increase in real estate taxes over a base year, attempt to limit those taxes for which the company will become responsible. For example, attempt to exclude responsibility for any assessments against the property or for taxes based upon the rental income of the landlord. If the lease provides that the tenant is to pay the increase in real estate taxes over a base year, make sure that the base year is defined as the first tax year after the building is at least 80% occupied. (The assessed valuation of the property, and consequently the taxes and the tax base, are artificially low during the base year due to the relationship between the assessment procedure and the "lease up" period.)

Repairs and Maintenance

In negotiating the repair and maintenance clause, try to specifically define those areas which will be the company's responsibility to maintain. The lease should then provide that everything else is the responsibility of the landlord. The lease should also provide that, in the event the landlord does not maintain and repair those items which are his responsibility, the company may do the required work and deduct or "offset" the cost of such maintenance from its rent. In addition, the lease should specify in detail what cleaning and janitorial services the landlord is to provide (such as during which days of the week) and types of service (e.g., emptying wastebaskets, vacuuming, cleaning walls, washing windows, changing light bulbs, cleaning bathrooms, etc.). Ask the landlord how much of the rent payment is applied to cleaning costs. Try to secure the right to receive the cleaning amount as a credit against the rent if the company wishes to contract for cleaning independently. The landlord may or may not agree with this approach. It is nonetheless a useful device to be assured of good cleaning results.

Utilities

Most leases provide that the landlord pays for all water and sewer charges, and heating and cooling costs as well. Electric and all other utilities consumed directly within the leased space will probably be paid for directly or indirectly by the company as tenant. If the lease contains a clause stating that the tenant shall pay any increase in utility costs over the base year, make certain that the base year is the first year after the building is at least 80% occupied. In addition, if the tenant is to pay for any utilities, try to get a separate meter installed, or try to have the company's costs established based on surveys by independent engineers. If the tenant is to pay for utilities based on the number of square feet contained in its premises compared to the number of square feet in the entire building, make sure that there is no very heavy user of electricity or fuel in the building. If there is, then the company's prorated share should be adjusted to take into account the heavy user. Try to include a clause that will exclude any heavy utility user from pro-rated utility charges.

Alterations

Try to reserve the right to make any alterations or additions to the interior of the premises so long as they do not affect the structural integrity of the building. In addition, avoid obligating the company to making any alterations to the premises that might be required by governmental authorities.

Hold Harmless and Indemnity Clause

As tenant, try to limit the company's responsibility for indemnity to only those liabilities arising out of the negligence or willful misconduct of the company or its employees. Make certain that the company is not responsible for indemnifying any liability arising from the landlord's negligence or willful misconduct.

If the company is to be charged with a pro rata share of the building's casualty insurance, try to make certain that the insurance covers only fire and extended coverage (and not expensive types of coverage such as flood and/or earthquake insurance). Again, if the tenant is required to pay increases in the cost of insurance over the base year costs, make certain that the base year is defined as a year when the building is at least 80% leased.

Destruction Clause

Insist that in the event of damage to the building by fire or other casualty, if the premises are not rebuilt within a certain period of time (for example, 90 days), the company has the right to cancel the lease. Try to provide that, during the rebuilding or restoration period, the rent will be abated in proportion to the extent that the premises cannot be used. Remember, the landlord should have "loss of rents" coverage, which would kick in if any tenant abates rent as a result of damage.

Insist that the landlord be obligated to repair or rebuild the premises if they are only partially destroyed (for example, less than fifty percent destroyed).

Condemnation Clause

In the condemnation clause (if there is one in the lease), insist that the company receive that portion of the award attributable to:

- Improvements paid for by tenant
- Tenant's trade fixtures and equipment
- Moving and relocation expenses of tenant
- The value of the tenant's leasehold estate

In addition, insist that, in the event of a partial condemnation which substantially alters the nature or effectiveness of the building or the company's business operations, the company should have the right to terminate the lease.

It is also important to have the right to terminate the lease if the number of parking spaces condemned is sufficient to create a hardship for the company.

Subordination

If the lease contains a subordination clause, the company, as tenant, should insist that it not be obligated to execute or subordinate its leasehold interest to the lien of any mortgage, unless the mortgagee agrees in writing that the lease will not be terminated in the event of a foreclosure as long as the company is not in default under the lease.

Assignment or Subletting

Attempt to obtain an assignment and subletting clause that provides as much freedom as possible to assign or sublet the premises. At a minimum, the landlord should be required not to unreasonably withhold consent to an assignment to subletting.

Notice Provision

A notice provision is very important, as it allows the landlord and tenant to know to whom and in what manner they must give notice to each other. An adequate clause would read something like the following:

"Any notice which may be or is required to be given by either party to the other hereunder shall be in writing and shall be deemed to have been fully given when personally served or when 48 hours have elapsed from the time when such notice was deposited in the United States Mail, certified or registered, and postage prepaid, and addressed as follows: to lessee at:

(Address),

or to such other place as lessee may from time to time designate in a notice to lessor; to lessor at:

(Address),

or to such other place as lessor may from time to time designate in a notice to lessee."

Negotiating Subleases on Expansion Space as Landlord

The following indicates the various subleasing arrangements that will be most favorable to the company, when it functions as a landlord.

As was the case in negotiating as a tenant, the principal goal in lease negotiations is to shift to the other party the risk of having to pay additional monies in connection with the lease resulting from the occurrence of unexpected events. Again, because that is the goal, almost all so-called "legal" issues are really "business" issues; that is, they involve the attempt to limit the responsibility of one party or the other to pay for unforeseen costs. This in no way suggests, however, that the company should not involve its attorney in developing the terms of the lease. The discussion that follows is written in business language; the company's actual lease will probably need to be written in legal terms. Therefore, the company's attorney should be involved in the negotiations and in the preparation or approval of lease language, as well as advising the company of local law and customs.

Parties

It is important to know exactly with whom one is dealing. Always try to obtain current financial statements of the potential tenant for the sublease. Also, make sure that the person or persons executing the sublease on behalf of the tenant have the proper authority, as well as the power to bind the tenant.

If the proposed tenant is a corporation, a corporate resolution of the tenant should be obtained stating that the sublease is authorized by the Board of Directors of the corporation, and that the persons who are signing the sublease on behalf of the corporation have the proper authority to do so. If the final financial statement of the corporation does not show substantial financial strength, then require that the corporation's stockholders either execute the lease as additional tenants, or execute a separate guarantee of the tenant's obligations.

If the proposed tenant is a partnership, obtain a copy of the partnership agreement or other document that states which persons are authorized to sign a sublease on behalf of the partnership.

If the proposed tenant is neither a corporation nor a partnership, then each of the individual owners of the company should be required to execute the sublease, as well as each of their respective spouses.

Premises

Make sure that the premises are clearly defined in the sublease. The sublease should state the number of rentable square feet for which you are charging the tenant, and it should clearly define the term *rentable square feet* (for example, does it include non-usable space such as bathrooms, elevator lobbies, utility rooms, loss factors, and the like?).

Improvements

If the company, as landlord, is going to construct improvements for the tenant, make certain that the plans and specifications for those improvements are worked out in detail before the sublease is signed, so that there will not be a dispute later over what, as landlord, was supposed to be provided for the tenant.

This part of the document will become an exhibit to the sublease and will be referred to as the "Work Letter." The terms of this Work Letter must be completely defined and reduced to specific quantities, dollar allowances and qualities before the lease can be executed. Detailed plans should be prepared and approved in writing by the company, as landlord, and the tenant. Under no circumstances can a lease be executed on the basis that these matters can be worked out after the sublease is signed.

If the company is going to grant to the tenant rights to use property outside the actual premises, such as the right to park in parking spaces, make sure that such an arrangement does not conflict with rights previously granted to other tenants. For example, if another tenant has been previously granted the right to use all parking spaces on a nonexclusive basis, it is inappropriate to grant another tenant the right to use certain parking spaces on an exclusive basis. In addition, make sure that definitions of the premises and rights granted to the tenant do not exceed the rights the company has under its own lease.

Term

As landlord, attempt to provide a firm date on which the sublease term will commence, but at the same time provide that the company will not be liable in the event that it is not able to deliver the premises to the tenant on the specified date. In addition, always specify a fixed sublease term. Resist giving the tenant options to renew or extend the sublease term. Options exist only for the benefit of the tenant, and therefore should be resisted if at all possible.

Obviously, it is important to make certain that the sub-lease term does not extend beyond the term the company has under its own master lease.

Rent

Always make sure that any increases in costs which the company, as landlord, must pay in connection with the premises are passed on to the tenant through appropriate escalation provisions. In addition, if the sublease term is a long one (that is, over five years), or if there are options to renew, then insist that the basic rent clause contain an escalation provision tied to the company's actual operating cost increases. Insist that the rent clause provide that the rent be paid without deduction or offset.

Tax Clause

Always attempt to have the tenant pay all real estate taxes levied against the premises, or at least insist that the tenant pay any increases in real estate taxes over a base year. Also make sure the definition of real estate taxes is as inclusive as possible, that is, it should also include all general and special assessments against the premises and taxes based on the company's rental income derived from the lease. Most particularly, it should cover any taxes for which the company is charged.

Repairs and Maintenance

In negotiating a repair and maintenance clause, try to define specifically those areas which are the company's responsibility to maintain as landlord. The sublease should then provide that everything else within the "premises" is the responsibility of the tenant. The sublease should also provide that, in the event the tenant does not maintain or repair those items which are his responsibility, the company may perform the required maintenance and recover the cost of such maintenance from that tenant.

Alterations

Attempt to restrict the right of the tenant to make alterations or additions to the premises. In no event should the tenant be allowed to make alterations or additions to the premises which affect the structural integrity or aesthetics of the building. Reserve the right to approve the tenant's plans for any alterations or improvements.

The sublease should contain a provision that gives the company or the building landlord the option to either:

- Require that the tenant remove any alterations or additions at the end of the term, at its expense, or;
- Allow the landlord to keep the alterations or additions to the premises.

If the tenant is required or allowed to remove the alterations or additions to the premises, he should be required to repair any damage caused by the removal of the alteration or addition. Try to see that the tenant is obligated to make any alterations or additions to the premises that are required by government authorities because of tenant's use of the premises.

Hold Harmless and Indemnity Clause

The sublease should contain a provision stating that the tenant indemnifies, defends, and holds the landlord harmless from any and all liabilities arising out of the use, condition, or occupancy of the premises.

Insurance

The sublease should require that the tenant provide comprehensive liability insurance covering liabilities related to his use, condition, or occupancy of the premises, and that such liability insurance should name the company as landlord as an additional insured. In addition, the liability insurance should contain a provision that it may not be cancelled without at least 20 days' prior written notice to the landlord and that the insurance is primary, and any insurance carried by the company and/or its landlord are in excess thereto.

The tenant should be required to provide certificates of insurance evidencing liability insurance coverage.

The sublease should also provide that the tenant will pay any increase in the cost of the fire insurance over and above the cost of the fire insurance during the base year and passed through by the company's landlord. As was the case with taxes, make sure the company recovers its costs.

Destruction Clause

The sublease should provide that, in the event of any destruction by fire or other casualty, the company as landlord has the option to either keep the sublease in effect and rebuild the premises, or terminate the sublease. Do not give the tenant the option to cancel the sublease. Insist that there will be no rental abatement in the event that the company elects to rebuild the premises; or the company should carry "loss of rents" coverage.

Condemnation Clause

In the condemnation clause, insist that the company receive the entire award except only for that portion attributable to a tenant's trade fixtures and equipment, and moving and relocation expenses. In no case should the company as landlord agree that the tenant receive any portion of the award attributable to the value of the tenant's leasehold estate or the so-called *bonus value* of the sublease.

Subordination

Make sure that the sublease contains a subordination clause that provides that the sublease will be, at the company's option, either prior or subordinate to the lien of any mortgage. The clause should also provide that the tenant, at the company's request, will execute any documents necessary to subordinate the tenant's leasehold interest to the lien of any mortgage.

Assignment and Subletting

The sublease should restrict the tenant's right to assign or sublet the premises without the company's consent. In addition, the sublease should provide that no involuntary assignment or assignment by operation of law will be valid without the company's consent.

Notice Provision

A notice provision is very important, as it allows the landlord and tenant to know to whom and in what manner they must give notice to each other. An adequate clause would read something like the following:

"Any notice which may be or is required to be given by either party to the other hereunder shall be in writing and shall be deemed to have been fully given when personally served or when 48 hours have elapsed from the time when such notice was deposited in the United States Mail, certified or registered, and postage prepaid, and addressed as follows; to lessee at:

(Address),

or to such other place as lessee may from time to time designate in a notice to lessor; to lessor at:

(Address),

or to such other place as lessor may from time to time designate in a notice to lessee. Lessee hereby appoints as its agent to receive the service of all dispossessors or restraint proceedings and notices thereunder the person in charge of or occupying the premises at the time, and if no person shall be in charge of or occupying the premises at the time, then such service may be made by attaching the same on the main entrance of the premises."

Making the Final Building Selection

Clearly, the task of selecting a new site for a facility and negotiating a final lease for that site is immensely complex. Work closely with the company's architect or space planner and attorney to compare the three, four, or five prospective buildings and control buildings which have been investigated in depth. For ease of comparison, summarize the various offers received according to each of the following:

- Rentable square footage required to supply usable area needs;
- Areas available in the future, either as option space or by right of first refusal;
- Total monthly rental costs and cost per usable square foot;
- Preferred length of lease;
- Building standard improvements and the relative value of each;
- Lease offers and how they relate to the preferred terms;
- Any other special incentives offered by a landlord;
- The physical assets and defects of each building;
- A subjective evaluation.

To properly evaluate the various lease offers under consideration, update the budget to indicate the impact that each lease offer would have on bottom-line costs. For instance, if one building is offering a greater dollar value in tenant improvement allowances, this will have an effect on the total budget and therefore on considerations relating to that particular offer. To do this, refer to the budget that was started during the evaluation process described in Chapter 1. Also review Chapter 4 as it relates to the preparation of a detailed budget estimate. Adjust the estimate to account for the new and more detailed information uncovered while evaluating available facilities. Indicate the dollar differential impact that each lease offer would have on the company's ultimate costs.

At this point, the facility or project manager should have developed a solid comparison between prospective buildings as to lease terms, tenant improvement letters, building shell space, floorplate efficiencies, rental costs per square foot of occupied space, building qualities and amenities. This information, together with the work completed in the previous chapter, provides the foundation needed to refine the company's actual needs.

Chapter 3

PROGRAMMING TO DETERMINE ACTUAL SPACE NEEDS

Introduction

Offices are, to a large extent, a reflection of the services, products and production techniques of the company the staff serves. Offices grew out of our collective need to know – to know what the products are that we are producing as represented by forms, agreements, drawings and specifications, etc.; to keep track of inventories, including what materials or parts were purchased, used, replaced and paid for; to keep track of sales, invoicing, accounts receivable, assets, liabilities, profits and losses; and to schedule production, pay employees and pay the proper amounts of taxes. In a very real sense, the office was an administrative, engineering and clerical image of the company's product lines. Offices received, correlated, and produced information. As long as a company produced the same product lines using the same techniques, materials, and labor, and sold to essentially the same customers or markets, there was little pressure to change. As products and services became obsolete through creative market pressures, however, office functions began to change to meet the new information demands. New devices and techniques in the form of technological advances were developed to assist in the process of receiving, analyzing and disseminating the new information. New ways of organizing workers were required to cope with the technology. New product development, technology utilization, and organizational management have been accelerating at an ever increasing pace.

To properly design for offices today, one must be not only aware of the state of the art in technologies, but also knowledgeable regarding the trends and objectives of the company and its philosophy toward meeting constantly shifting goals. Most importantly, the designer must plan not only for currently defined functions, but also for the undefined. As it is clearly not possible to create specific designs for that which is undefined, the planner or designer must plan for maximum *flexibility* and *adaptability* in order that the various components of the design standards may be modified to serve the new demands.

Establishing Space Standards and Basic Design Elements

The establishment of space standards and other basic design elements is a necessary pre-condition to the interior design process. However, before selecting space standards, it is valuable to consider the function of those standards.

A workstation provides a specific place for a person to work and communicate. It should be designed to provide a comfortable and useful space in which to perform the desired function. Workstation standardization is an attempt to provide a universality of workstation design to accommodate a variety of work functions. These may seem like conflicting objectives. However, it fits today's direction—away from repetitive job function, and toward team or unit organization; away from clerical staff or support workers to multi-disciplined, knowledge workers.

The overall design concept should recognize that an individual may be functioning as a team member on one or more teams, while still retaining a primary job description. For example, an engineer may be working with one or a number of team-driven projects in different portions of a building, while still working on a special engineering design solution, unrelated to any specific team project. The project manager's attempts to provide for the requirements of these multi-faceted work environments (workstations) will be dictated by the nature of the company's own culture and the philosophies of senior management. Even though management may recognize these trends, most offices are still designed as physical places to deal with a work process, with only a secondary regard for the needs of the people performing the work.

The designer of a set of workstation standards must now design for a wide variety of job functions within any "standardized" workstation design. The problem is that the trend toward greater and greater flexibility and toward satisfying variable job functions, may not be clearly described today. In fact, those job functions may not even exist today. Yet the designer (and the standards) must provide pragmatic solutions, now. Recent studies indicate that the "churn rate" in many of our larger offices is about 25% per year. It is possible that from the time a design is fixed until a space is occupied due to reassignments, resignations or terminations may actually result in a churn and turn-over rate approaching 50%.

For the facility manager or designer to be able to not merely respond to changing operating techniques and technologies, but to anticipate them, they must be aware of changing styles of management. As we enter a phase in our business life in which greater emphasis is on product development, refinement or redevelopment, with its trend toward a unit or team approach, we may require project-oriented spaces, perhaps with speciality stations such as a CADD station, a PC station, a research station, small or large group meeting areas, and spaces for just plain thinking. These all might be clustered in such a manner as to provide maximum collaboration between team members, and at the same time, individualized acoustical, lighting and temperature/humidity environments.

Quantitative and Qualitative Requirements

Solutions to the conflicting demands for standards with maximum flexibility and specific accommodations can be met by approaching the problem from two directions. Both approaches are based on an assessment of corporate culture—both current and future—that is to say how the company conducts its business now, and how it can be anticipated it will be conducted in the future.

Before defining space standards, carefully consider the "given":

- What are the interior modules of the building for which space plans are being prepared?
- What is the core to exterior wall dimension? What are the column bay spacings?
- How much exterior or natural light is available?
- What are the travel distances likely to be (is the building long and narrow, or short and square)? For example, the analysis for a particular building may suggest that a 9' module is superior to a 10' module without adversely impacting a workstation function or design.

After the analysis of the given building configuration, the next step is to select a standard of dimensional modules. The most popular modules appear to be 3', 4' (or 2'), and 5' (or 30") modules. The criteria for the selection of one module over another should be based on a broad assessment of the tasks now being performed by the company's various employees, what equipment they need, and what kind of conditions they should have in which to perform their jobs. We should recognize that as people become multi-tasking, they require more equipment and therefore more space in which to accommodate that equipment. Design analysis of the company's principal buildings, especially bay spacing or exterior wall-to-core depths, may be a deciding influence on which module is selected. Whatever criteria is used, it is important to settle on one module for a particular building.

Example

By establishing one consistent dimension of, say, 9' (using the 3' module), one can generate a set of about 6 reasonably flexible workstation sizes, all with one or two common dimensions as shown in Figures 3.1 and 3.2.

An alternative consistent dimension might be 10' (using the 5' module). One can generate a set of flexible workstation sizes.

Figure 3.1

9' x 6' =	54 SF
9' x 9' =	81 SF
9' x 12' =	108 SF
9' x 15' =	135 SF

and

12' x 12' =	144 SF
12' x 15' =	180 SF

Figure 3.2

10' x 7.5' =	75 SF
10' x 10' =	100 SF
10' x 12' =	120 SF
10' x 15' =	150 SF
10' x 20' =	200 SF

The second element is to settle on no more than two or three alternate manufacturers of systems furniture for a particular building or complex. Recognize that most manufacturers can supply a number of different systems furniture lines. Try to narrow the selections down to no more than one line per manufacturer, and make sure that the lines selected are as comparable to one another as possible. This will permit maximum recycling of systems furniture and components as needs and arrangements change.

Limiting modules and size standards to the greatest degree possible is quite important. This approach may not be the most economical from a pure space utilization viewpoint, but it permits the greatest flexibility in both the re-configuration of workstations and the supplying of environmental elements—ambient lighting, HVAC systems and communications cabling. Surveys of some major companies indicate that space standards, especially regarding size, frequently range from four to seven or eight in number.

Further on, several basic job functions/workstations used in offices today will be discussed and illustrated. Some of the workstations are of the same dimension. Some are for the same function, but demonstrate different solutions. As an appropriate module has been defined, the configuration of the workstation becomes much more simple and flexible.

Some examples of auxiliary areas, such as conference rooms and reception areas, have also been included. They are not necessarily standardized, but can be standardized at this point in your planning. To the extent practicable, these areas should be designed to the same module standard as are the workstations, for maximum future flexibility.

The descriptions explain not only what each item is, but why it was chosen, where it is to be used, when it needs to be ordered, and how it is to be installed. There are some special pointers which should be of value when making design decisions.

Workstations

Workstations consist of systems furniture, a modular system of panels, work surfaces and other components (files, shelves, etc.) which, when assembled, provide the user with an appropriate work space and equipment necessary to perform his or her particular job tasks. A workstation can be designed for each job category within a company, including clerical, production, technical, supervisory, and managerial personnel.

A workstation can be a system of low and/or mid-height wall panels with fully integrated work surfaces, drawer units, file units, flipper door cabinets, work organizers, etc. (all, or most all, manufactured by the same company) which are modular in nature and interchangeable.

A workstation can be a system of panels assembled to create partial height walls, with or without electrification, in which conventional furniture is placed. This configuration is quite popular, as some facilities managers feel that it provides simpler and quicker reconfiguration by permitting an individual worker's desk to be relocated from one workstation site to another. An alternative is to provide each person with a personal pedestal component which can be wheeled from one workstation to another. This is particularly useful for team configurations when a person may wish to move from a PC station to a CADD station, and perhaps to a research station.

A workstation can also be just a desk and a chair sitting in the middle of the floor. (A receptionist's workstation is often configured in this way.) It can also be a conventional office with four walls, with conventional office furniture.

Where Workstations Are Used
Open-plan office workstations are used in 60% to 85% of a typical U.S. corporation's facilities.

Selection Criteria
Because of its extreme flexibility, systems furniture can be used in almost any department of a facility. Its components and panel heights may be varied as needed to accommodate different functions, and the work being performed.

Workstation configurations can be changed as a company's job function requirements change. If systems furniture panels are used, it is possible to get systems with electrical and communications raceways integrated into the panels. This allows all power, telephone, and data cables to be run through the panels instead of through the walls and floors, which are much less flexible. It is also important to look for a system that is well constructed; has a good, well designed variety of components; is reasonably priced; and is produced by a company that is well established in the industry and able to deliver an order on time and at the quoted price. Although most companies will work with "customer's own material" (COM), it is advantageous to deal with a supplier who has a good selection of fabrics, finishes and colors suitable to the company's own design standards.

Purchasing
Large companies often arrange to purchase directly from a manufacturer through a National Buying Contract. This means that the manufacturer has agreed to sell directly to the company at a fixed discount off list price. This sales price usually includes delivery and installation. More often, systems furniture is sold through commercial office furniture dealers. Dealers can often supply pricing as good as those under national contracts, and they have the advantage of being local businesses interested in servicing the company and supplying its continuing furnishing needs.

If the cost of new systems furniture is prohibitive, it is worth inquiring about reconditioned products. Many dealers and some manufacturers are able to supply reconditioned and guaranteed products.

Ordering
Methods of ordering and delivery depend on the quantity of goods involved. For a current delivery schedule, check with the manufacturer or dealer. Most manufacturers have a "Quick Ship" program for one or more of their common product lines. The response time for these products is one to six weeks for delivery, but rarely for large quantities, often not more than ten units. Quantities over this and for expanded color and fabric selections, the lead times will be up to 10 to 15 weeks.

Installation
Installation can be by the manufacturer using one of his local representatives, or by the local commercial office furniture dealer. The installers will set up and attach panels, work surfaces, and components as shown on the company's furniture plans. Some systems will have an electrical system pre-installed at the factory. Others will be shipped as components with the electrical wiring harnesses or cables to be installed by the installing contractor. Rarely, if ever, will the furniture installer make the final electrical connections to the building's electrical system or install the communications or data cables. The electrical contractor will typically make the final electrical connections.

Special Points
There are many variables in the features and construction of workstation furniture systems. It is, therefore, a good idea to first discuss appropriate systems with the company's interior designer or architect, and contact local

representatives or dealers for several manufacturers' systems. Ask colleagues from other companies or facilities who have workstation systems for comments on their system. Ask them if they would order the same system if they could start from the beginning. Ask for their recommendations as to the dealer they used. Keep in mind that no dealer carries more than three to four systems lines. Watch out for the line that is carried by only one dealer in the area, as it may not be possible to get a competitive price on it, without a national contract. It is possible to try negotiating a fixed mark-up fee with the dealer, but keep in mind that all dealers are not treated equally by the manufacturer. Unfortunately, this is true of some of the very best lines in the market. The best alternative is to bid one line against another.

Talk to two or three dealers. Ask for a presentation of their products. Eliminate all but the two or three best systems, then ask their rep or dealer to present a mock-up of one of your company's standard workstation configurations, or to take you to an actual facility using their product line so you can get a "feel" for how the product will look in your facility. Most dealers are willing to do this; all you have to do is ask. Once the preferred systems have been selected, ask the dealer or rep to help write the specifications for the specific system of choice. Often the dealer will be willing to provide you with furniture installation drawings and quantity takeoffs as a part of their service. This service is often an inclusive part of the dealer's markup.

Remember, installation of systems furniture requires the coordination of two to four other contractors (for the system itself, electrical, communications and often data), in order to make the workstation completely functional.

Be sure to review the company's designs with the local building authorities to determine what steps (if any) are required before beginning the installation. Although nearly all systems furniture manufacturers of electrical power components will be UL-recognized, this should be verified. It is also important to determine whether the local building department will allow the use of factory-produced electrical harnesses.

Once a particular manufacturer's product line has been chosen, that same product should be ordered, as needed, for future work so as to maintain continuity throughout the facility.

Sample Workstations

The following section lists a number of workstation configurations, with components. These illustrations (Figures 3.3 through 3.13) are generic. Some of these workstations may be similar to what the company is already using, while others may be entirely different. They are presented here as a starting point in the development of the company's own set of "workstation standards." While it is worth trying to keep the basic square footage modules within the company's selected standards, configurations can be modified to meet specific needs and/or restrictions. It is very important to either select some of the types presented here, or have a set designed, and to end up with from four to seven modules with one constant dimension. Within a given size and shape, there may be a rather extensive variety of configurations, depending on the job function. Use the ones presented here for ideas. Prepare and have approved your own standards.

Type A

Clerical/Administrative

47 SF Workstation
- Desk (work surface)
- Return
- Rear work surface
- 1 Pedestal with
  - 1 Box drawer with pencil tray
  - 1 Box drawer
  - 1 File drawer
- 2 Hanging shelves
- Task chair

Variables
- Desk
  - Right-hand return or
  - Left-hand return
  - Panel-supported or
  - Free-standing
- Chair
  - Frame color (finish)
  - Fabric
  - Style, color, COM
- Panels
  - Height
  - Width
  - Electrification
  - Finish
  - Color
  - Tackable
  - Hard surface (painted)
  - Acoustical
  - Fabric
  - Style, color, COM

Figure 3.3

Type B

Technical/Clerical/Specialist

49-64 SF Workstation

- Corner work surface
  – 2 Returns
- 2 Pedestals each with
  – 1 Box drawer with pencil tray
  – 1 Box drawer
  – 1 File drawer
- 2 Hanging shelves
- Swivel arm chair

Variables

- Desk
  – Right-hand return or
  – Left-hand return
  – Panel-supported or
  – Free-standing
- Chair
  – Frame color (finish)
  – Fabric
  – Style, color, COM
- Panels
  – Height
  – Width
  – Electrification
  – Finish
  – Color
  – Tackable
  – Hard surface (painted)
  – Acoustical
  – Fabric
  – Style, color, COM

Figure 3.4

Type C-1
Programmer/Supervisor/Technician

64 SF Workstation

- Corner work surface
  - 2 Returns
- 2 Pedestals each with
  - 1 Box drawer with pencil tray
  - 1 Box drawer
  - 1 File drawer
- 1 Hanging shelf
- 1 Hanging shelf with flipper door
- Guest Chair
- Swivel arm chair

Variables

- Desk
  - Right-hand return or
  - Left-hand return
  - Panel-supported or
  - Free-standing
- Chair
  - Frame color (finish)
  - Fabric
  - Style, color, COM
- Panels
  - Height
  - Width
  - Electrification
  - Finish
  - Color
  - Tackable
  - Hard surface (painted)
  - Acoustical
  - Fabric
  - Style, color, COM

Figure 3.5

Type C-2

Supervisory/Managerial

72 SF Workstation

- Desk (work surface)
    - Return
    - Credenza work surface
    - 1 Pedestal with
    - 1 Box drawer with pencil tray
    - 1 Box drawer
    - 1 File drawer
    - 2 Hanging shelves
    - 1 2-Drawer lateral file
- Swivel arm chairs
- 2 Guest chairs

Variables

- Desk
    - Right-hand return or
    - Left-hand return
    - Panel-supported or
    - Free-standing
- Chair
    - Frame color (finish)
    - Fabric
    - Style, color, COM
- Panels
    - Height
    - Width
    - Electrification
    - Finish
    - Color
    - Tackable
    - Hard surface (painted)
    - Acoustical
    - Fabric
    - Style, color, COM

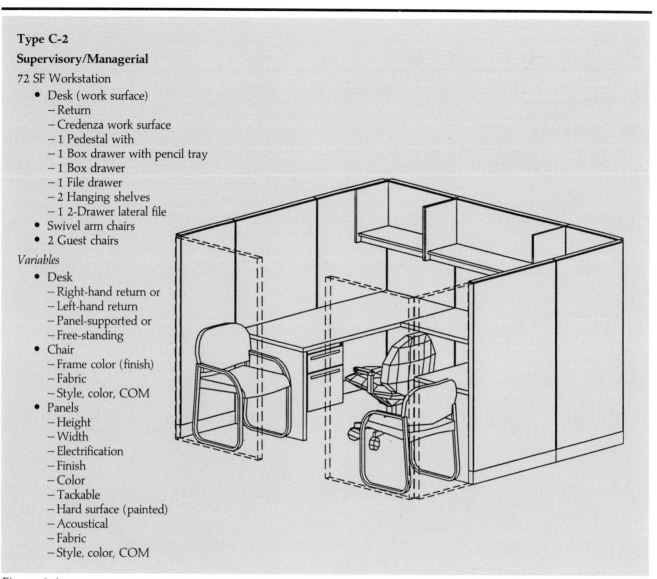

Figure 3.6

Type D

Supervisory/Managerial/Technical/Professional

125 SF Workstation

- 2 Corner work surface
  - 3 Returns
  - Work/conferencing table (work surface)
- 2 Pedestals each with
  - 1 Box drawer with pencil tray
  - 1 Box drawer
  - 1 File drawer
- 1 Hanging shelf
- 2 Hanging shelves with flipper doors
- 1 2-Drawer lateral file
- Swivel arm chair
- 2 Guest chairs

Variables

- Desk
  - Table-desk without drawers
- Paper organizers
- Task lighting
- Credenza
  - Storage cabinet
  - Drawer pedestal
- Chair
  - Frame color (finish)
  - Fabric
  - Style, color, COM
- Panels
  - Height
  - Width
  - Electrification
  - Finish
  - Color
  - Tackable
  - Hard surface (painted)
  - Acoustical
  - Fabric
  - Style, color, COM

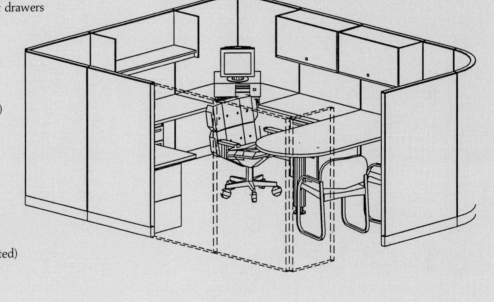

Figure 3.7

Type E-1

Supervisory/Group Leader/Technician

122 SF Workstation

- Desk: free-standing with double pedestals, each with:
  - 1 Box drawer with pencil tray
  - 1 File drawer
- 2 Hanging shelves with flipper doors
- Credenza with double pedestals each with
  - 1 Box drawer with pencil tray
  - 1 File drawer
- Swivel arm chair
- 2 Guest chairs

Variables

- Desk
  - Table desk without drawers
- Book shelves
- Paper organizers
- Task lighting
- Credenza
  - Storage cabinet
  - Drawer pedestal
- 2 Drawer lateral files
- Chair
  - Frame color (finish)
  - Fabric
  - Style, color, COM

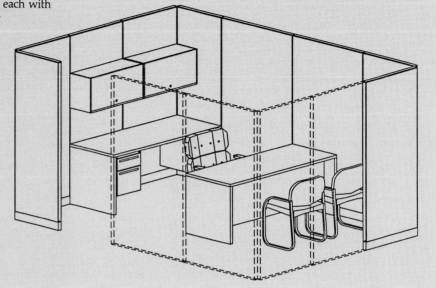

Figure 3.8

Type E-2

Supervisory/Group Leader/Technician

150 SF Workstation, Private Office

- Desk: free-standing with double pedestals, each with:
    - 1 Box drawer with pencil tray
    - 1 File drawer
- 2 Hanging shelves with flipper doors
- Credenza with double pedestals each with
    - 1 Box drawer with pencil tray
    - 1 File drawer
- Swivel arm chair
- 2 Guest chairs

Variables

- Desk (Table-desk without drawers)
- Book shelves
- Paper organizers
- Task lighting
- Credenza
    - Storage cabinet
    - Drawer pedestal
- 2 Drawer lateral files
- Chair
    - Frame color (finish)
    - Fabric
    - Style, color, COM

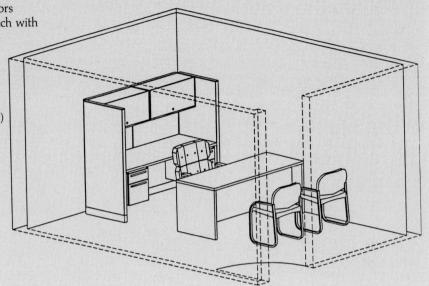

Figure 3.9

Type F-1

Supervisory/Managerial

160 SF Workstation

- Desk: free standing with double pedestals, each with
  - 1 Box drawer with pencil tray,
  - 1 File drawer
- 2 Hanging shelves with flipper doors
- Credenza with 1 2-drawer lateral file
- Swivel arm chair
- 4 Guest chairs

Variables

- Desk (Table-desk without drawers)
- Book shelves
- Paper organizers
- Task lighting
- Credenza
  - Storage cabinet
  - Drawer pedestal
- Chairs
  - Frame color (finish)
  - Fabric
  - Style, color, COM
- Panels
  - Height
  - Width
  - Electrification
  - Finish
  - Color
  - Tackable
  - Hard surface (painted)
  - Acoustical
  - Fabric
  - Style, color, COM

Figure 3.10

Type F-2

Supervisory/Managerial

160 SF Workstation, Private Office

- Desk: free-standing with 2 double pedestals, each with
  - 1 Box drawer with pencil tray
  - 1 Box drawer
  - 1 File drawer
- 2 Hanging shelves with flipper doors
- Credenza with
- Swivel arm chair
- 5 Guest chairs
- Side or small table

Variables

- Desk (Table desk without drawers)
- Book shelves
- Paper organizers
- Task lighting
- Credenza
  - Storage cabinet
  - Drawer pedestal
- Chairs
  - Frame color (finish)
  - Fabric
  - Style, color, COM
- Panels
  - Height
  - Width
  - Electrification
  - Finish
  - Color
  - Tackable
  - Hard surface (painted)
  - Acoustical
  - Fabric
  - Style, color, COM

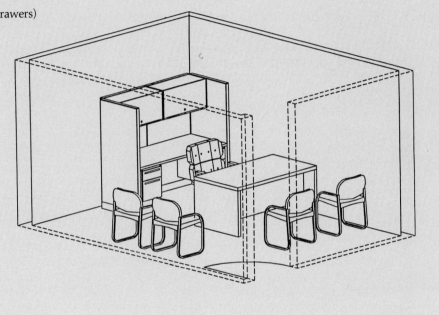

Figure 3.11

Type G-1

Division Manager/General Manager

Vice-President

255 SF Workstation, Private Office

- Desk: free-standing with double pedestal, each with
  - 1 Box drawer with pencil tray
  - 1 File drawer
- Credenza
- Double pedestal each with
  - 1 Box drawer with pencil tray
  - 1 Box drawer
  - 1 File drawer
- Swivel executive chair
- 5 Guest chairs
- Side or small table

Variables

- Desk (Table-desk without drawers)
- Bookshelves
- Paper organizers
- Task lighting
- Credenza
  - Storage cabinet
  - Drawer pedestal
  - 2 Drawer lateral files
- Chairs
  - Frame color (finish)
  - Fabric
  - Style, color, COM

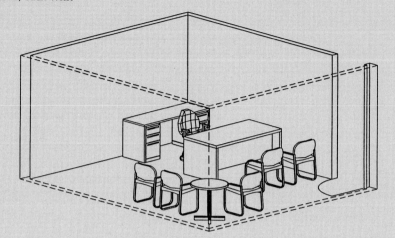

Figure 3.12

Type G-2

Division Manager/General Manager

Vice President

300 SF Workstation, Private Office

- Desk: free standing with double pedestal, each with
  - 1 Box drawer with pencil tray
  - 1 Box drawer
  - 1 File drawer
- Credenza double pedestal each with
  - 1 Box drawer with pencil tray
  - 1 File drawer
- Swivel executive chair
- 2 Guest chairs
- Conference table
- 4 Swivel arm chairs

Variables

- Desk (Table desk without drawers)
- Bookshelves
- Paper organizers
- Task lighting
- Credenza
  - Storage cabinet
  - Drawer pedestal
  - 2 Drawer lateral files
- Chairs
  - Frame color (finish)
  - Fabric
  - Style, color, COM

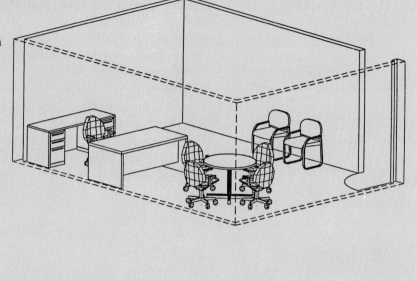

Figure 3.13

Executive Office Furniture

Where it Is Used
Executive office furniture will include desks, credenzas, chairs and tables especially selected for use in private offices for management personnel.

Selection criteria: Comfort, status or image, durability, quality, and conformity with overall design concepts.

Purchasing
Executive furniture is purchased from local contract furniture dealers. It can be bid or negotiated, depending on the quantities involved. Large amounts should be bid to two or more contract furniture dealers. For small amounts, negotiate to obtain the best pricing from a dealer with whom the company has a good working relationship. Do not accept substitutions for specific furniture lines listed in the company's design standards.

Ordering
Executive furniture has a long lead time due to the manufacturing time needed to produce high-quality items. Wood finish and custom upholstery takes 12-16 weeks or longer, if fabrics and materials are not in stock. Check with the manufacturer for their current delivery schedule.

Installation
The dealer should deliver and install all furniture for private offices.

Special Points
Private office areas require special attention by the project designer or architect.

Auxiliary Areas

Conference Room
There are two basic types of conference rooms shown herein; comfortable lounge-type seating groups, and pull-up or swivel-tilt conference-type seating with a table. The quality of the elements can vary greatly, from vinyl-covered chairs and plastic laminate tables, to leather seating with fine wood or stone custom conference tables.

Where Furnishings Are Used
Lounge chairs are used in enclosed lounge conference rooms or areas designed for confidential meetings, especially by personnel who do not have private offices. This type of conference room or area is frequently overlooked, yet it can be one of the most useful.

The more conventional conference room or area typically will use average-quality chairs and tables for "work" conference rooms, centrally located near the work areas of departments. Conference rooms are often shared by departments. Their locations should be convenient to all sharing groups.

High-quality conference seating and tables are used in formal conference rooms located in areas easily accessible to management and visitors.

Selection Criteria
The important criteria for conference room furnishings are: comfort, durability, and ease of movement.

Purchasing
Conference room furniture can be purchased through local commercial office furniture dealers. Dealers can often supply pricing as competitive as those under national contracts, and they have the advantage of being local businesses interested in servicing local clients and supplying continuing furnishing needs.

Ordering

Conference furniture should be ordered 12-15 weeks prior to planned move-in. Check with the manufacturer for a current delivery schedule.

Installation

All conference furniture is installed by the furniture dealer. Any assembly and/or adjustments are handled at that time.

Special Points

The following section is a listing of some typical conference room furniture requirements with illustrations (Figures 3.14 through 3.18). Again, these are for reference only and not intended to address the exact needs of a specific company or facility.

Conference Rooms

Four-Person Open

133 SF
- 4 Lounge chairs
- Side or coffee table

Variables
- Chairs:
  - Frame color
  - Fabric
  - Style, color, COM

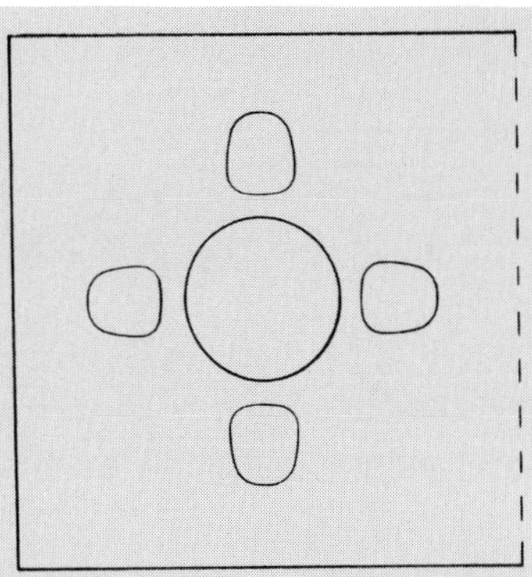

Figure 3.14

Four-Person Open Lounge-Type

133 SF
- 4 Arm (Conference) chairs
- 42" or 48" diameter conference table

Variables
- Dri-erase marking board
- Table
  - Round
  - Square
  - Finish
- Chairs
  - Frame color
  - Fabric
  - Style, color, COM

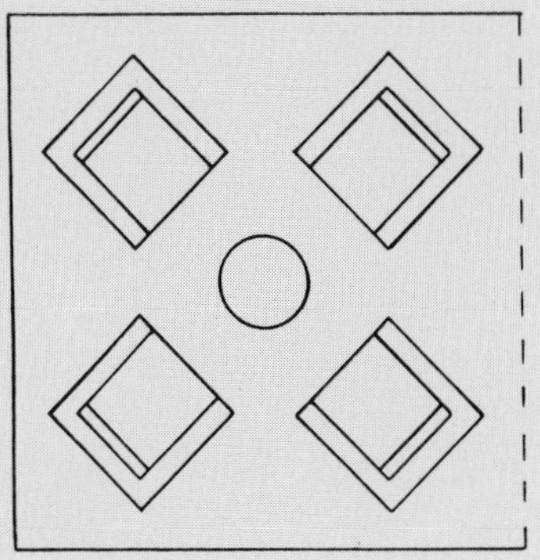

Figure 3.15

Six-Person

144 SF
- 6 Arm (Conference) chairs
- 42" or 48" diameter conference table
- Side table

Variables
- Dri-erase marking board
- Table
  - Round
  - Square
  - Finish
- Chairs
  - Frame color
  - Fabric
  - Style, color, COM

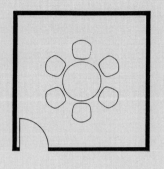

Figure 3.16

Eight-Person

216 SF
- 8 Arm (Conference) chairs
- 42" or 48" x 96" conference table
- Side table

Variables
- Dri-erase marking board
- Table
  - Boat shaped
  - Rectangular
  - Finish
- Chairs
  - Frame color
  - Fabric
  - Style, color, COM

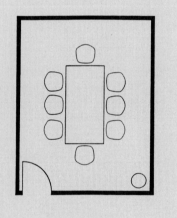

Figure 3.17

Twelve-Person

297 SF
- 12 Arm (Conference) chairs
- 42" or 48" x 144" conference table
- Side table
- Built-in server and wall cabinets, or large credenza/server

Variables
- Dri-erase marking board
- Table
  - Boat-shaped
  - Rectangular
  - Finish
- Chairs
  - Frame color
  - Fabric
  - Style, color, COM

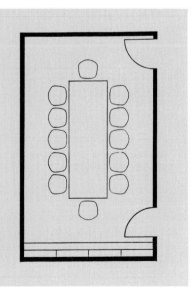

Figure 3.18

Training Room Furniture

Generally, folding seminar-style tables, stacking chairs and lecterns will suffice as equipment for training rooms. If the company's training rooms involve data entry or other forms of computer or terminal stations, or other forms of speciality equipment, it may be necessary to have these facilities especially designed and to select training station furniture, equipment and seating.

Special audiovisual equipment may be needed for this facility. This is a highly complex and rapidly changing area of involvement, and it may be necessary to retain a consultant in this field. Not only can the consultant specify the exact equipment that will be needed, but he or she can also direct you to the various source-vendors and dealers. Keep in mind the frequency or lack of frequency of use of some of the more costly equipment. It is often easily rented. Also keep in mind the requirements for safeguarding this type of equipment. Not only is it fragile and often difficult to operate, but is highly subject to theft. If possible, construction of a special projection room might be considered. Such an area could keep the equipment safe from theft or damage, while also ensuring that it is handy for use.

Where it Is Used

In education and training facilities.

Selection Criteria

Furniture is easily moved or shifted into different arrangements as needed (e.g., seminar style, auditorium style), and easily stored, except for the specialty training facilities mentioned above.

Purchasing

Local commercial office furniture dealers can often supply pricing as good as those under national contracts, and they have the advantage of being local businesses, interested in service and a continuing opportunity to supply future furnishing needs.

Installation

Installation is normally performed by the furniture dealer, using the company's own furniture plans.

In the following section, there are listings and illustrations (Figures 3.19 through 3.21) of typical training rooms.

> *Note: The square footage given here are illustrative only. Actual educational and training room sizes depend on the size of the facility they serve and on the type of training room configuration required. However, training rooms will run from 12 SF to 25 SF per seat. These examples show three configurations from among the many possibilities. You should carefully consider the types of educational requirements before laying out the specific rooms.*

Auditorium Style

594 SF

- 46 Seats
- 13 SF/seat

Figure 3.19

Seminar Style

725 SF
- 36 Seats

20 SF seat
- Stacking chairs
- Folding tables
- Projection table
- Small side table
- Podium
- Chalkboard
- Projection screen
- Dri-erase marking board
- Audiovisual equipment

Variables
- Chair frames
  - Finish
  - Style
  - Color
  - Table sizes

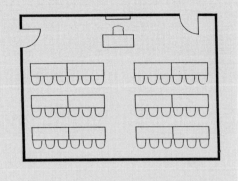

Figure 3.20

Education and Training Room
- Audiovisual style 12-seat

280 SF
- 12 Seats
- 23 SF seat (with reverse seating arrangement)
- Stacking chairs
- Folding tables
- Small side table
- Podium
- Chalkboard
- Projection screen
- Dri-erase marking board
- Audiovisual equipment

Variables
- Chair frames
  - Finish
  - Style
  - Color
  - Table sizes

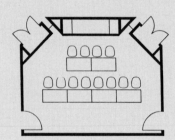

Figure 3.21

Lunchrooms

Lunchrooms, cafeterias, and dining rooms are a significant subject for review and discussion. The choices a company makes in dealing with this subject can be costly, relatively difficult to modify (because of the lack of flexibility inherent in kitchen facilities), and very important to all personnel. In many cases, cafeteria facilities have been designed too small. There also appears to be a strong preference for freshly prepared (on site or commissaried) food, versus vending machine service. There continues to be a need for some private dining rooms for conference-type meetings, but there is also a trend toward the elimination of an "officers' dinning room" or special eating areas.

The question of food subsidy is of concern during the design phase due to the issue of access control to prevent uninvited nonemployees from using a company cafeteria and enjoying company-subsidized meals. Access control will also impact the number of "turns" (number of times each seat is "turned" over to a new customer/user), for which the seating areas are designed.

When considering the design direction for the company's cafeteria, it is important to consider which of two design directions it may wish to follow or may find more appropriate. These are:

- a quiet, subdued environment designed to relax the employees, which might be appropriate if the majority of users work in high stress job, such as customer relations, as would be found in a claims processing facility.
- an upscale, exciting environment designed to stimulate, as might be desirable for actuarial or statistical analysts.

Obviously there are many levels in between these two extremes. The decision will likely be predicated on corporate culture—and the major job functions the employees are performing.

Have in mind that no matter how well the furniture may be arranged in the eating areas, those people using the facility will very likely try to rearrange it to their own wants. They will often prepare special arrangements to accommodate special groups of friends and associates, or simply pull together tables to accommodate a larger lunch group.

When considering various furniture lines and designs, keep in mind that this furniture will take a considerable beating, especially carpeting, chair arms, upholstery, table bases, table tops and edges. Do not be led into selections chosen primarily because they "look nice." Check with a knowledgeable dealer to determine if the selected items have a good history of durability and are easily cleaned.

Consider four-leg chairs or sled-based chairs, strong enough to resist breaking when leaned back in. Consider the use of carpet tiles laid in checker-board, herringbone, striped, or similar patterns for flexible replacement. Resist solid colors both in seating fabrics and carpeting due to their tendency to make food stains more obvious. A non-monolithic carpet tile installation will not show the effects of cleaning, wear and fading, whereas a solid field tile or roll carpet will. Remember that attic stock carpeting, whether roll or tile, will be a different color due to fading of the carpet that is in service, even if the carpet has been in service for only a short time.

Lunchroom Furniture

Lunchroom furniture consists primarily of tables, chairs, and stools. Another option for this area of the building is comfortable lounge seating.

Where it Is Used

In lunchrooms and cafeterias.

Selection Criteria

Look for durability, comfort, ease of maintenance, and appropriateness of design. Plastic laminate tables with wood or soft plastic edging are best. Avoid self-edged tables as they are likely to chip and delaminate. Table bases should be heavy and stable, with adjustable glides. Avoid painted bases unless you wish to achieve a "worried" look. Seating should have nontilt rear legs, and upholstered seats should wrap around the base and not have spaces in which food can be trapped. Wood frames should be stained, not painted. Most paint will chip. Avoid chairs whose arms are likely to scrape on the under edge of the tables.

Purchasing

From local furniture dealers. (See previous section on Executive Office Furniture.)

Ordering

Allow 12 to 16 weeks before the desired delivery date.

Installation

By the furniture contractor according to locations shown on the company's furniture plan for the area.

Special Points

Lunchroom furniture should be selected with two purposes in mind: to offer a change of environment from the normal work areas, and to accommodate the special maintenance problems associated with food service.

Reception Room Furniture

This category includes desks and chairs, lounge seating, coffee and side tables.

Where it Is Used

In all reception areas.

Selection Criteria

Because the reception area gives visitors their first impression of a facility, this furniture should be chosen with particular care as to design and finish. Both the design and furniture selection for each reception room should set the theme for what is to take place inside of the particular office being served by that reception area. For example, personnel interviewing requires a different reception psychology than that desired for the executive area. In addition, items should be chosen for their durability, comfort, ease of maintenance, and appropriateness of design in relation to the other design elements.

Purchasing

By various local furniture dealers. (See Executive Office Furniture, earlier in this chapter.

Ordering

Allow 12 to 18 weeks from placement of order to the date delivery is required.

Installation

By the furniture contractor according to locations shown on the company's furniture plan.

Special Points

The reception desk should be placed so that the receptionist faces visitors as he or she works, without having to turn around. If space allows, divide reception room furniture into groups or separate seating areas, so that groups of visitors may have a sense of privacy while they wait for their meeting.

If the receptionist is performing any form of work other than answering the phone and greeting/directing visitors, the receptionist desk may have a gallery rail or transaction top to shield the work surface and computer screen from the view of visitors.

Note: The plans in Figures 3.22–3.26 are examples only. Many different configurations are possible within the same square footage. The components of each are listed below.

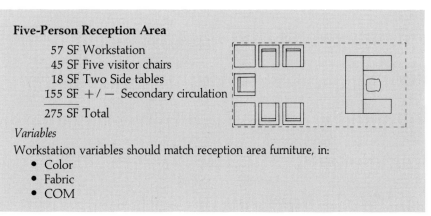

Four-Person Reception Area

57 SF One workstation
36 SF Four visitor chairs
9 SF Side table
148 SF +/− Secondary circulation

250 SF Total

Variables

Workstation variables should match reception area furniture, in:
- Color
- Fabric
- COM

Figure 3.22

Five-Person Reception Area

57 SF Workstation
45 SF Five visitor chairs
18 SF Two Side tables
155 SF +/− Secondary circulation

275 SF Total

Variables

Workstation variables should match reception area furniture, in:
- Color
- Fabric
- COM

Figure 3.23

Six-Person Reception Area

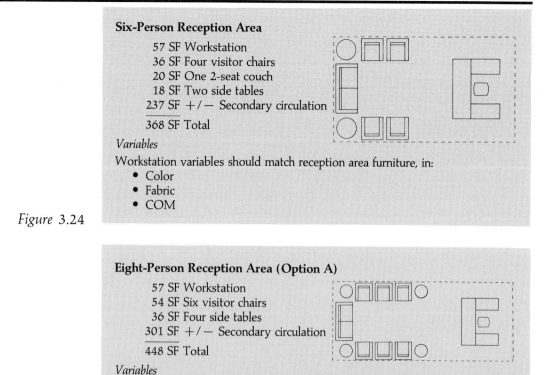

57 SF Workstation
36 SF Four visitor chairs
20 SF One 2-seat couch
18 SF Two side tables
237 SF +/− Secondary circulation
—————
368 SF Total

Variables

Workstation variables should match reception area furniture, in:
- Color
- Fabric
- COM

Figure 3.24

Eight-Person Reception Area (Option A)

57 SF Workstation
54 SF Six visitor chairs
36 SF Four side tables
301 SF +/− Secondary circulation
—————
448 SF Total

Variables

Workstation variables should match reception area furniture, in:
- Color
- Fabric
- COM

Figure 3.25

Eight-Person Reception Area (Option B)

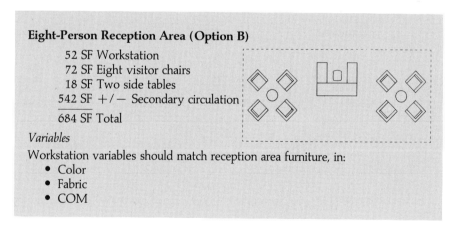

52 SF Workstation
72 SF Eight visitor chairs
18 SF Two side tables
542 SF +/− Secondary circulation
—————
684 SF Total

Variables

Workstation variables should match reception area furniture, in:
- Color
- Fabric
- COM

Figure 3.26

Accessories

Accessories consist mostly of desktop items such as letter trays, pencil cups, calendars, and memo holders, along with other items such as wastebaskets, coffee servers and trays. There are two kinds of accessories, operational and executive. Operational accessories are typically made of special plastic or metal, formulated to be scratch- and stain-resistant. Executive accessories can be made of more exotic materials.

Where They Are Used

Accessories are used in individual workstations, executive offices, training rooms, visitors' offices and conference rooms. It would be wise to develop standard accessory "packages" for each job type or workstation standard in the company. These standards can then be consulted for correct allocations. Using a standardized accessories package carefully tailored to the job function will prevent constant re-evaluation of needs for every new employee.

Purchasing

Some of these products are purchased from the company's commercial furniture dealer based on specifications developed by the project or facilities manager, or the company's designer or architect. Some products can be purchased from office supply dealers. Small items like accessories are sold individually, usually with a price break for large quantities (e.g., 10-24 items = first discount break; 25-96 items = second discount break; 96 and more items = best price).

Ordering

Accessories are stocked in small quantities and can usually be received in about 1-4 weeks. Large quantities can take up to 12 weeks. Check with the company's dealer.

Installation

If purchased through the company's commercial furniture dealer, delivery and set-up may be performed by them. If purchased directly or through an office supply outlet, delivery will usually be by UPS or other freight company direct. The company distributes internally.

Special Points

Watch for quantity discounts as well as minimum order quantities when re-ordering. Check lead times and availability.

Plants and Interior Landscaping

The placement of plants, whether live or artificial, and of varying types and species, can serve a number of functions besides providing a natural feeling in a rigid and structured environment. Plants can be used to designate a traffic pattern, or a change of direction, or to signal an obstruction. They can define a department or area without the use of walls. NASA studies indicate that live plants help to reduce indoor air pollution and provide oxygen.

Where They Are Used

Plants are typically located in the open-plan operational areas, with the remainder in private offices, conference rooms, and reception areas. In smaller companies the majority of plants are to be found in the private offices rather than in the open-plan or public areas.

Selection Criteria

Live plants add a fresh, natural element to the interior working environment, but do require maintenance and therefore present an operating expense. Plastic or silk plants are easier to maintain and require little operating expense. Select whichever feature is more important to the company. A mixture of real and artificial plants may be a good choice, based on criteria such as vulnerability to cold, accessibility for watering, maintenance costs, the amount of direct sun, or the desirability to establish a special reception room environment.

Purchasing

Live plants are usually leased from and maintained by a local nursery, with maintenance and plant rental fees billed as a monthly charge. The best pricing can be obtained by bidding (see "Plants and Containers," in Chapter 15.) It may be advantageous to purchase the company's plants. Some prefer to buy two sets of some plants and use a service to rotate and maintain them. Leased plants are also often changed periodically for revitalization or variety. Plastic or silk plants can generally also be obtained from a nursery.

Ordering

Plants are usually available locally and can be obtained relatively quickly. Allow approximately 3-4 weeks in the order for assembly and delivery.

Installation

The local nursery will deliver all plant materials to their proper location as shown on the plant location plan (see "Construction Documents," in Chapter 7.)

Special Points

Consult local nurseries for recommended plant materials. The availability of particular plants and the lighting conditions of the region will dictate which species are best suited for a facility.

Art

Art involves decorative graphics or free-standing sculpture in the office areas. The program may include: prints/lithographs, photography, tapestries, weaving, paintings and sculpture, or any number of art forms.

Where it Is Used

Art may be used throughout the office facility, primarily in the public and common use areas, elevator lobbies, reception spaces, major circulation areas, lunchroom lounge areas, in enclosed conference rooms, offices and special areas. Its purpose is to create visual interest, provide color stimulus, and add life to the working environment. Refer to the section entitled "Art," for information on evaluating program requirements, appropriate locations, scale, size, quantities, security and budgets (Chapter 16).

Purchasing

Art may be purchased outright, leased, or leased with a return option before the final purchase is negotiated. It is recommended that, for a program involving more than a few pieces, an art dealer or art consultant be retained.

Ordering

Program requirements, budget allocations and consultant selections should be approved three months prior to the anticipated move-in date to allow adequate time for program analysis, budget reviews, selections, approvals, framing, and installation.

Installation

The approved pieces will be installed by the art consultant or dealer in accordance with the art program.

Special Points

The art program should be tailored to enhance the environment of the facility. Art forms selected should be in keeping with the design philosophy being followed in the new facility. The selection of art forms allows a variety of options. Art programs should be tailored not only to enhance the environment by adding visual interest, but to stimulate staff interest in the expressive art forms currently being used by artists at the local, regional, and/or national levels.

Graphics & Signage

The graphics program should include whatever signage is required to establish the company's corporate identity, provide parking information, and convey location information for the various subsidiaries, departments, and special common use areas housed within the company's facilities.

Where it Is Used

Signage is used in the parking and entry areas, attached to the building exterior, or applied to the exterior walls and doors as applicable. Signage also includes such items as exterior and interior corporate signage, building identification, lobby directories, subsidiary and departmental suite entry signage, and directional information as required to properly guide and assist employees and the public who use the facility, as well as individual employee identification.

Purchasing

Signage must be manufactured in accordance with the specific program requirements for the facility. (Refer to Chapter 14.) Consult with the company's designer for information on qualified suppliers.

Ordering

Manufacturing, delivery and installation of signage varies somewhat depending on the suppliers' manufacturing demands, but generally requires 10-12 weeks from the date of order. Allow 4-6 weeks prior to order entry for the development of budgets and approvals of the facility's specific requirements.

Installation

Signage is usually delivered and installed by the manufacturer or its authorized local distributors.

Special Points

Review all developer stipulations and local signage ordinances to ensure conformity before submitting a program for approval and order entry. Review the Americans With Disabilities Act for various requirements regarding graphics and signage.

Developing Space Requirements Questionnaire

Determining the total amount of space required for a facility should occur in two ways. The first is from the top down. This evaluation will be predicated on past historical growth patterns in product sales, production or new product development as related to office employee growth. In reviewing these patterns, one should look at the company from its vertical organization — that is senior management, middle management and staff growths — and from its horizontal organization — divisions, product groups, service departments, etc. Some of these divisions may have only existed for a short time, while others may have ceased to exist. The objective is to establish a mathematical "smoothing" to account for any increase or decrease anomalies. By examining these projections and having them analyzed and reviewed by senior management (including the CEO, Vice President of Finance, Vice Presidents of Sales and Marketing and Director of Strategic Planning), a consensus of future personnel growth should begin to emerge. There are consulting companies with very sophisticated measuring systems for performing these analyses. These services are quite valuable and should be considered for large projects. The resulting data should describe population trends by division or group, and management levels within the divisions.

Once this phase of the evaluation has been defined, the next step is to evaluate from the staff level up. During this evaluation, an attempt will be made to develop finite growth estimates, department by department. Operational problems will also be investigated – if they might affect how the space is to be used. During this phase, it is helpful to make frequent comparisons with the historic review.

To help find out more precisely how much space the company will require, both in a year or so (the change date), and projected out for five years from the change date, a set of forms is provided in the Master Forms Section at the end of the book. These forms can be assembled into a questionnaire for the project or facility manager or department heads to complete. The questionnaire covers basic departmental staffing, work flow, equipment needs, and special requirements.

However, before photocopying and using these questionnaire pages, the company's own Space Standards should be developed, to the extent possible. Review the forms before copying and distributing them. Questions that are not relevant to a particular company can simply be ignored. Other areas of inquiry may not be covered by these forms. If so, either modify the appropriate form or prepare a new form to cover the questions that should be addressed. Once these objectives have been accomplished, proceed with the review process.

Workstation and Grade Level Standards should be completed at this time. It will then be possible to complete those portions of the questionnaire that deal with workstation type and area. The final workstation configuration does not need to be determined at this time. That is more appropriately left to the Design Development phase of the project.

The space selection and leasing process should also have progressed to a point at which a final determination can be made of the module and workstation standards best suited to the specific building and space intended for occupancy. The final selection process should be based on the following criteria:

- the various work functions to be performed
- the department sizes and anticipated change volatility
- the size of the building space or floor plates
- the core-to-exterior wall depths
- the column bay spacing
- any other influencing modules

From the standards analysis, six to eight different workstation standards will likely be generated, covering the majority of the company's needs. If the design modules, as recommended earlier in this chapter, have been completed, there may be as few as five or six basic sizes, with many configurations within in each size module.

There is one more step to be performed prior to being able to complete the questionnaire. That is to assign each individual his or her appropriate workstation. To do this, it will be necessary to create some form of workstation-type assignment based on an approved corporate policy. The more common of these assignment criteria are: job descriptions, job titles, salary levels, etc. We have prepared a typical workstation assignment matrix based on the workstation areas reviewed earlier in this chapter. Remember it is neither necessary nor desirable to have designed the specific configurations at this point. Our example (Figure 3.27) does include a Furniture Code designation which will not be needed at this point. It is included in order to complete the Grade Level package. This example is a modified version of one used by a major casualty insurance company. It outlines a typical format for which job levels might receive which workstation types.

GRADE LEVEL	JOB TITLE	WORK-STATION TYPE	OFFICE TYPE	NOTES	S.F. AREA	FURNITURE CODE
STANDARDS – OFFICE PLANNING						
1- 8	Clerical/ Administrative	A	Open	-	47	D
1- 8	Technical/ Clerical/ Specialist	A or B	Open	1	49	D
9-13	Programer/ Supervisor/ Technician	C-1	Open	1	64	D
9-13	Supervisor/ Manager	C-2	Open	1	72	D
14-17	Supervisor/ Manager/ Technician/ Professional	D	Open	1	125	D
14-17	Supervisor/ Group Leader Technician	E-1	Open	1,2,3	122	D or C
14-17	Supervisor/ Group Leader Technician	E-2	Private	1,2,3	150	A or B
18-20	Director/ Division Manager/ Vice President	F-1	Open	1,2,3	160	D or C
18-20	Director/ Division Manager/ Vice President	F-2	Private	1,3	160	B
21-24	Division Manager/ General Manager/ Vice President	G-1	Private	1,3	255	A
21-24	Division Manager/ General Manager/ Vice President	G-2	Private	1,3	300	A

Figure 3.27

Notes

1. Workstation type and configuration to be determined by analyzing the work performed by the individual involved.
2. Open office for nonofficer; private office for officers.
3. Furniture for secretarial/reception area to have the same code as the office it serves.

In Addition:

Individuals with a requirement for confidentiality should receive a private office commensurate with their grade level, except that no office shall be smaller than a size "E."

The prestige of certain positions will require the use of a "functional override."

Functional Override

Certain job functions, due to their range of activities, the need for more prestigious space, additional privacy or other functional requirements, may demand special consideration in workstation design. For example:

- A regional Vice President should be upgraded to a "G" from an "F."
- A Grade 17 Group Leader who heads a division should be upgraded to an "F" workstation from an "E."

Authority to reassign workstations requires the authority of the Officer in Charge.

Furniture Selection Codes

A. Wood Furniture — Approved design from standards; couch, club chairs, side tables, coffee tables, bookcases, lamps and other accessories as appropriate to the needs of the user; wall coverings, art, special trim, etc. as appropriate to overall design, with direction and approval from user.

B. Wood Furniture — Approved design from standards; accessories and equipment as needed to fill functional requirements; some latitude in wall color, art and miscellaneous items from alternates provided by designer of specific location.

C. Free-standing — Approved design from standards; type of equipment based on functional requirements; design to conform to system used below; acoustical and visual privacy achieved through use of systems panels.

D. Systems Furniture — Workstations and components based on functional requirements.

Figure 3.27 (continued)

It is important to not only develop the company's own corporate design policy statement and standards, but also to try to cover the above-mentioned areas. Edit them to fit the company's needs, but by all means, be certain to include all the above concerns. At a minimum, the standards should include a statement of policy, drawings of approved workstations, meeting rooms, etc., and guidelines as to how they are to be selected.

An alternative to handing out the questionnaires for department heads to complete, is an interview procedure directly with the department heads. The questionnaire forms can be used as a guideline of questions to ask, and the answers can be filled out on the forms during an interview process. An added benefit to interviews is the likelihood of clarifying or filling in information not covered by the questionnaires. Experience has found this to be a more effective strategy. Keep in mind that senior management may know of future plans which may affect a department in a manner contrary to the information supplied by the department head. Have each department head review the workstations information before they assign workstation types to each staff position.

To get maximum benefits from this process, the person who has been put in charge of the questionnaire process should follow these steps:

Step 1: Distribute one copy of the questionnaire to each department head. Note that, before it can be distributed, the first page of each questionnaire should be filled in to indicate the change date year, and the year five years from that date. Also fill in the name and telephone number of the person the respondent should call if he or she has any questions about how to fill out the questionnaire.

Meet with the department heads either in a group (or groups) or individually. Explain to them that the company may be involved in a potential redesign and relocation of their departments, not just for current needs, but for the way the department head sees them five to six years into the future. The need for supporting data (for their projections) as well as qualitative assessments of each job function should be stressed. Be sure all department heads understand that they must answer the questions with the utmost care, as their answers will shape future planning efforts. Emphasize that departmental personnel projections must take into consideration both industry and management philosophy regarding growth trends and market projections.

Figures 3.28 through 3.32 are sample questionnaire forms.

Space Requirements Survey

Facility: _____ Location: _____

Department: _____ Date: _____

Prepared By: _____ Phone: _____

- Read instructions at top of each sheet carefully before completing each page.

- Please type or print all answers.

- The term "section" refers to those people who report to you.

- When estimating projections, the number in each column, (P) Present, (M) Move-In and (F) Future, should represent a cumulative total.

- For all totals of personnel, furniture and equipment:

   A. Under "Present," include all AUTHORIZED staff positions, whether currently filled or not.

   B. Under "Move-In," indicate your cumulative projections for personnel and equipment.

   C. Under "Future," indicate your cumulative projections for personnel and equipment.

- Please keep in mind that this information is necessarily preliminary. Your responses will be discussed with you later, and additional detail will be taken at that time.

- All information will be updated and refined as the time approaches to develop final layouts and plans.

- Additional Information

_____

_____

_____

_____

_____

_____

- Questions regarding these forms and this process may be directed to:

Name: _____ Phone : _____

_____

_____

Figure 3.28

Adjacencies: External and General Data *1 of 1*

Facility: _____   Location: _____

Department: _____   Date: _____

Prepared by: _____   Phone: _____

The following questions explore the working relationship
of your section to other sections within your department
or facility.

1. What hours of the day and what days of the week does your section work?

_____

_____

_____

2. Does your section fluctuate in size on a seasonal or project basis? If so, please describe.

_____

_____

_____

3. If your section is currently dispersed, what efficiencies in staff or equipment could be
   realized through consolidation?

_____

_____

_____

List, in descending order of importance, those sections and facilities which your section
should be located near in order to operate effectively. Briefly explain your reasons. Indicate
by code whether the adjacency is: (E) Essential; (I) Important; or (C) Convenient

Section or Department	Explanation	Code

Figure 3.29

Adjacencies: Internal

Facility: _____ Location: _____

Department: _____ Date: _____

Prepared By: _____ Phone: _____

The purpose of the following questions is to acquaint us with your section and to give us a general understanding of the work that is performed by your personnel, and to indicate the most logical arrangement of people and equipment within your section.

- Briefly summarize the overall function of your section. _____

_____

_____

_____

_____

- Briefly describe or diagram the flow through your section. _____

_____

_____

_____

_____

- What would be the ideal adjacencies, proximity, and working relationships within your section?

_____

_____

_____

_____

_____

Figure 3.30

Workstations

Facility: _____ Location: _____

Department: _____ Date: _____

Prepared By: _____ Phone : _____

- List all authorized staff positions in your section, whether presently filled or not.

- Indicate the number of males and females in each position. (This may become important in determining restroom needs.)

- Indicate the total number of people in each position at present and your projections for cumulative number at move-in date and at a future date.

- Group all positions according to desired internal adjacency requirements, such as "supervisor", followed by people supervised.

- Select the appropriate workstation for each staff position based on job function.

Staff Position	M	F	Present	Move	Future	Workstation		SF Area		
						Type	Area	Present	Move	Future
Totals										

Figure 3.31

Equipment Work Areas

Facility: _____   Location: _____

Department: _____   Date: _____

Prepared By: _____   Phone: _____

- Using the descriptions of equipment types accompanying this form, list additional equipment your department or section requires.

- List only those items which are NOT included within individual workstations or private offices. (Chairs, work surfaces, desks, etc. are normally part of a workstation).

- List all items which take up floor space in your section. Do not list tabletop equipment such as typewriters, fax machines, and VDT terminals unless they require their own stands.

Equipment	Number Required			Type	Size			Space Required	Area		
	Present	Move	Future		W	D	H		Present	Move	Future
A. Vertical Files Ltr.											
B. Vertical Files Lgl.											
C. Lateral Files											
D. Book Shelves											
E. Cabinets											
F. Counter Tops											
G. Built-In Drawers											
H. Tables											
I. Copy Machines											
J. Printers											
K. Other Equipment											
L. Special Furniture											
Totals											

Figure 3.32

Reviewing and Tabulating Questionnaires

Step 2: To ensure that special management policy directions affecting staffing are taken into account, have all completed questionnaires reviewed by the manager or officer in charge of each department. As mentioned above, senior management may be aware of undisclosed policies or strategic plans which may have substantial effects on the information collected.

Note: During this review process, check to be sure that department heads and/or branch managers have allowed floor space to accommodate the career training, seasonal positions, and special consultant accommodations which some departments may require (e.g., auditors visiting the accounting department).

Step 3: Total the various personnel projections and compare these against available history of past growth.

Step 4: Verify final personnel projections with the head of the facility. (From time to time, a department may make an unrealistic projection of personnel requirements. It is most important for facility heads to evaluate and validate every personnel projection.)

Step 5: Verify that each staff position has been assigned a workstation type, and that the workstation is indeed suitable for that job category. The results can be verified with the company's human resources department. If there are discrepancies, the decision of the department head should be accepted until a review by senior management.

Revising Completed Survey Results

When all of the questionnaire forms have been completed, the information contained on them can be reviewed. Follow a step-by-step approach to this process.

General Review

Review each form, department by department, for completeness and reasonableness. If a piece of data is unclear, take the time to review it with the department head. A significant problem may be uncovered, which can be dealt with during redesign. Simple modifications to certain workstations, auxiliary areas or adjacencies may present themselves, with unexpected benefits. Or, a simple error may be discovered, with major facility consequences.

Senior Management Review

After reviewing all the initial data and making all indicated corrections, take the results to senior management for their review. Attempt to do this on a department by department basis. Try to work directly with senior management. It is very likely that some departmental information will receive management-directed modifications. Modifications will have to be done carefully and certain departments may have to be revisited to secure additional or modified information based on meetings with senior management. Make any final changes to the data and begin final tabulating.

Interpreting Space Needs

Based on final work space assignments, fill in the "area" portions of each department's "workstations" section of the questionnaire as shown in Exhibit 3.31 and add them up to obtain a subtotal in "assignable square feet" for workstations.

Example

The "Hard Work" Department of a large company says it has one superintendent now, will still have one superintendent when it moves in, and expects to have only one superintendent five years after the change date. The superintendent is to occupy workstation E. The form is completed by first determining that workstation E-2 takes up 122 SF of area; then calculating the square footage required using the following formula:

(number of positions) x (workstation area)

$$1 \quad \times \quad 122 \text{ SF} \quad = 122 \text{ SF}$$

Since there will be no future change in the number of superintendents, write 150 in the shaded spaces under "Pres.," "Move," "Fut."

All of the Hard Work Department's responses to the workstation questions are entered in on the form in Figure 3.33. The correct area allocations are noted in handwriting in the spaces provided.

Step 6: Now determine how much space will be required to accommodate each department's equipment needs and auxiliary areas. To do this, carefully evaluate the requirements reported in the questionnaires.

Equipment and Work Areas

Evaluate estimated present, move-in, and future needs and apply a square footage requirement to each item listed. (See the Equipment type forms.)

- Standard files (Equipment types A and B) and lateral files (Equipment type C):
  - When under 10 in quantity, allow 10 SF each.
  - When 10 or over, allow 7.5 SF each.

Note: Verify listed filing requirements to be sure that reported quantities and file types are consistent with the company's standards for filing methods. A filing survey may be required.

- Bookshelves (Equipment type D) and Cabinets (Equipment type E):
  - When under 10 in quantity, allow 13.5 SF each.
  - When 10 or over, allow 10 SF each.
- Countertops (Equipment type F):
  - 10 SF per lineal foot.
- Built-in Drawers (Equipment type G):
  - No allowance necessary.
- Tables (Equipment type H):
  - 35 SF each.

Other Special Equipment (such as a free-standing copier) must be calculated by noting its actual dimensions, calculating the floor space it occupies, then doubling that area (to account for secondary circulation and necessary clearances for collators and servicing — check manufacturers' technical data).

Auxiliary Areas

Reception Areas

Evaluate all reception area requirements listed to determine the possibility of sharing among departments. This can be done by considering preliminary departmental adjacencies before assigning area allocations. To determine space requirements after allowing for shared use, simply multiply the amount of seating needed by 25 SF per person.

Work Areas

Verify all requests for special work areas, as requests are often based on existing conditions, and may not take into account new equipment and techniques. Apply square footage numbers only after this evaluation is complete.

Note: General storage rooms are often overlooked. Consult knowledgeable staff members or the company's outside consultant to verify the impact of special programs which may affect storage areas.

Workstations

Facility: **Branch Alpha Division** Location: **Fairview, IL**

Department: **Hardwork** Date: **11/22/93**

Prepared By: **David Johns** Phone: **555-8989**

- List all authorized staff positions in your section, whether presently filled or not.

- Indicate the number of males and females in each position. (This may become important in determining restroom needs.)

- Indicate the total number of people in each position at present and your projections for cumulative number at move-in date and at a future date.

- Group all positions according to desired internal adjacency requirements, such as "supervisor", followed by people supervised.

- Select the appropriate workstation for each staff position based on job function.

Staff Position	M	F	Present	Move	Future	Workstation Type	Area	SF Area Present	Move	Future
Superintendent		X	1	1	1	E-1	122	122	122	122
Assistant Superintendent	X		1	1	1	E-1	122	122	122	122
Office Supervisor		X	1	1	1	C-2	72	72	72	72
Asst. Office Supervisor		X	1	1	1	C-2	72	72	72	72
District Facility Manager	X		1	2	2	C-1	64	64	128	128
Temp. Entry clerk	–	–	0	3	3	A	47	0	141	141
Entry Control	–	–	0	2	3	B	49	0	98	147
Totals	2	3	5	11	12			462	755	804

Figure 3.33

Conference Rooms

As with reception areas, carefully evaluate the potential for sharing conference rooms. To determine how much space is required after allowing for shared use, simply multiply the amount of seating needed by 25 SF per person.

Other Facilities

This section of the Auxiliary Areas page allows room for information that may not have been covered by any other section of the questionnaire. Again, evaluate each item listed here with great care before applying appropriate square footages.

Example

The Hard Work Department's responses to the Equipment Work Areas and Auxiliary Areas of the questionnaire are shown in Figures 3.34 and 3.35. The filled-in information was provided by the department.

Step 7: Now total each department's space needs as calculated in Step 5 (workstations) and Step 6 (equipment and auxiliary areas), and add 15%, or slightly more, to allow for adequate circulation between workstations, offices, desks, around files, and support equipment within a department, plus about 10% to 15% to get around between departments.

The Hard Work Department's total space requirements are summarized on the form in Figure 3.36.

Step 8: Next, run a cumulative subtotal of all departments.

Example

The example in Figures 3.37a and 3.37b presents the total area required by all departments in a sample division of a large company. (Hard Work is one of these departments.) As can be seen, this sample branch office will require a total of 56,847 SF to satisfy the anticipated future departmental needs.

Step 9: Now that the space requirements of the facility's departments are known, the space needed for common use areas not specified as a part of the departmental questionnaires should be identified. Typical branch common areas (not already covered under "Auxiliary Areas") might include space for a lunchroom, a lounge, and education and training areas. Larger facilities might require, in addition, a cafeteria (instead of a lunchroom), private dining areas, a full commercial kitchen, medical facilities, computer rooms, tape libraries, exercise and possibly recreational facilities.

Each facility's requirements for these areas will vary greatly, depending on the availability of off-site facilities or whether there are facilities already existing in the new building itself. Before applying any numbers here, carefully analyze the company's needs and existing availabilities, and be certain to take all options into account.

After verifying the need, if numbers are to be applied, use the following guidelines. For lunchroom and lounge areas: multiply the total personnel by 4.5 SF per person. For education and training: multiply the total personnel by 3.75 SF per person. For assistance in estimating special common area requirements, consult the company's in-house expert or outside consultant.

Step 10: Once the facility's total space requirements for common use areas have been determined, add the same 25% to 28% circulation factor as used in Step 7.

Step 11: Next, add the total area requirements for departments (see Step 8) to the total obtained in Step 10. This sum represents the total usable square footage the facility will require.

Equipment Work Areas

Facility: **Branch Alpha Division** Location: **Fairview, IL**

Department: **Hardwork** Date: **11/22/93**

Prepared By: **David Johns** Phone: **555-8989**

- Using the descriptions of equipment types accompanying this form, list additional equipment your department or section requires.

- List only those items which are NOT included within individual workstations or private offices. (Chairs, work surfaces, desks, etc. are normally part of a workstation).

- List all items which take up floor space in your section. Do not list tabletop equipment such as typewriters, fax machines, and VDT terminals unless they require their own stands.

Equipment	Number Required			Type	Size			Space Required	Area		
	Present	Move	Future		W	D	H		Present	Move	Future
A. Vertical Files Ltr.	18	18	18	A-5				7.5	135	135	135
B. Vertical Files Lgl.	1	1	1	B-5				10	10	10	10
C. Lateral Files											
D. Book Shelves											
E. Cabinets	2	3	3	E-3				10	20	30	30
F. Counter Tops											
G. Built-In Drawers											
H. Tables	5	3	3					35	175	105	105
I. Copy Machines											
J. Printers											
K. Other Equipment											
L. Special Furniture											
Tables- Folding	1	2	2					10	10	20	20
Lateral Files Locking	0	2	3					10	0	20	30
Totals									350	320	330

Figure 3.34

Auxiliary Areas

Facility: Branch Alpha Division   **Location:** Fairview, IL

Department: Hardwork   **Date:** 11/22/93

Prepared By: David Johns   **Phone:** 555-8989

Indicate only those areas that are used by your section or
department, or are shared with one or two other sections.
If more than one work area or conference room is required,
list in "other facilities" and indicate size.   P-Present; M-Move-In; F-Future

Reception/Waiting Area

			Area
Assume this is a general reception area on your floor, does your Section require a separate waiting/ reception area.	(A.) No, not at all. B. Yes, one for us. C. Yes, we can share.	P	
		M	
If you checked B or C, specify the number of seats required. Seats X 25 SF = Area	Present _____ Seats Move-in _____ Seats Future _____ Seats	F	

Work Area

			Area
Does your section require a separate work area, i. e. library, collating area, etc.	A. No, not at all. (B.) Yes, one for us. C. Yes, we can share.	P	150
		M	150
If you checked B or C, specify the number of square feet required.		F	150

Conference Area

			Area
Does your section require a separate conference room ? (check at right)	A. No, not at all. B. Yes, one for us. (C.) Yes, we can share.	P	100
		M	200
If you checked B or C, specify the number of seats required. Seats X 25 SF = Area	Present __4__ Seats Move-in __8__ Seats Future __8__ Seats	F	200

Other Facilities (specify)

	Area		Total	
	P		P	250
	M		M	350
	F		F	350

Figure 3.35

Square Foot Summary

Facility: __Branch Alpha Division__   Location: __Fairview, IL__

Department: __Hardwork__   Date: __11/22/93__

Prepared By: __David Johns__   Phone: __555-8989__

		Present	Move-In	Future
Personnel count	M	2	5	5
	F	3	6	7
Area summary for:		Square Feet		
		Present	Move-In	Future
Workstations		452	755	804
Equipment		350	320	330
Auxiliary areas		250	350	350
1.				
2.				
3.				
4.				
5.				
Shared functions				
1. Small conference room - 4 seat - Enclosed		100	100	100
2.				
3.				
4.				
5.				
Subtotal		1152	1525	1584
Plus circulation factor (**27** %) (about 28%)		311	412	428
Total		1463	1937	2012

Figure 3.36

Area Requirements by Department

No. _____

Facility: __Branch Alpha Division__   Location: __Fairview, IL__

Prepared By: __David Johns__   Phone: __555-8989__ Date: __11/22/93__

Departments	Personnel			Area Requirements		
	* Present	** Move	*** Future	* Present	** Move	*** Future
Administration	6	8	8	1320	1594	1594
Quality Control	10	13	13	2100	2713	2713
Marketing	20	24	26	4040	4900	5725
Legal	6	8	8	1910	2494	2494
Transcription	6	8	8	932	1219	1219
Alpha Group	5	11	12	1463	1937	2012
Sales	20	25	30	3500	4375	5288
Admin. for Office Services	2	2	2	300	300	300
Data Entry	6	7	7	900	977	977
Bravo Division	16	18	18	3009	3210	3581
Data Processing	4	4	4	4804	5655	5655
Central Files	10	10	14	3050	3050	3950
Mail + Supplies	3	3	3	1810	1395	1935
Audit	4	5	5	901	1012	1012

* PRESENT footage is a total of the space that would be required for present personnel, + special areas, + a circulation factor.

** MOVE-IN footage is based on a total of the space requirements for projected personnel as of the move-in date, + minimal expansion, + special areas , + a circulation factor.

*** FUTURE footage is based on a total of the space requirements for projected personnel (5) years after move-in, + special areas, + a circulation factor.

NOTE: All of the above represent needs exclusive of common areas such as lunch rooms, training facilities, etc. and are in usable square feet rather than rentable square feet.

Figure 3.37a

Area Requirements by Department

No. _____

Facility: __Branch Alpha Division__   Location: __Fairview, IL__

Prepared By: __David Johns__   Phone: __555-8989__   Date: __11/22/93__

Departments	Personnel			Area Requirements		
	* Present	** Move	*** Future	* Present	** Move	*** Future
Education + Training	12	4	4	1540	2030	2030
Reproduction	—	—	—	716	874	938
Production	17	18	19	3400	3800	4138
Lunchroom + Lounge	—	—	—	0	2250	2250
Hardwork	5	11	12	1463	1937	2012
Strategic Planning	8	9	9	1440	1831	1831
Engineering	11	13	13	3300	3818	3818
Customer Service	5	7	7	940	1375	1375
Totals	176	208	222	42,838	52,746	56,847

* PRESENT footage is a total of the space that would be required for present personnel, + special areas, + a circulation factor.

** MOVE-IN footage is based on a total of the space requirements for projected personnel as of the move-in date, + minimal expansion, + special areas , + a circulation factor.

*** FUTURE footage is based on a total of the space requirements for projected personnel (5) years after move-in, + special areas, + a circulation factor.

NOTE: All of the above represent needs exclusive of common areas such as lunch rooms, training facilities, etc. and are in usable square feet rather than rentable square feet.

Figure 3.37b

Example

In the case of the sample branch facility, five divisions will share the space. The example in Step 8 showed how the area requirements of one division were derived. (Similar calculations had to be made for each of the other four divisions before this stage could be completed.)

As the chart in Figure 3.38 shows, the five divisions need several common use areas. The total usable square footage required by the entire facility five years after the change date is 129,749 SF. The branch manager of the sample branch office will have to lease enough space to accommodate these needs.

Evaluating Existing Furniture

It is often necessary or desirable to use some or all of the company's existing furniture in a new facility. To help decide how much and which items can be used, and where it should be located, the following inventory process can be used. It consists of a Standard Inventory Form (on which code number, quantity description, condition, dimensions, present and new location, and mover tag number of each item of furniture is entered.) See the Master Forms Section for a blank form. A reduced-size sample appears in Figure 3.39.

After a thorough review of the present facility, follow the steps outlined below.

Step 1: Take a Physical Inventory. The most efficient approach is to walk through each floor (or room) from one end to the other, listing each item. It is helpful to take instant-developing photographs of each type of furniture, and each one-of-a-kind item. In the course of the inventory process, duplicate items will be encountered. Repeat photographs for items that are exactly the same do not have to be made, but keep a running count of each piece of furniture and note that count on the inventory form. Obtain the following information for each individual piece of furniture and list it on the inventory form:

- Dimensions: height, width, depth, left- or right-hand return, base height, return height, etc.
- Finish: Wood type, color, finish (lacquer, oil, etc.). Metal type, color, finish (matte or gloss).
- Upholstered Pieces: color, material, pattern, texture, and manufacturer's name (if possible).
- Features: Any unusual features or variations on a standard piece of furniture. Also, whether or not the hardware is standard.
- Condition: Assign each piece one of these classifications: G (GOOD), F (FAIR), P (POOR), D (DAMAGED). (Describe the type of damage.)
- Location: Present location: establish a method of identification (sticker, tag, etc.) for inventoried pieces.

Step 2: Assign Furniture Codes. Organize the inventoried items by assigning furniture codes as in the following example. Figure 3.40 shows an expanded code listing.

Facility Area Requirements

Facility: Branch   **Location:** Fairview, IL

Prepared By: David Johns   **Phone:** 555-8989   **Date:** 11/22/93

Divisions	Personnel			Area Requirements		
	* Present	** Move	*** Future	* Present	** Move	*** Future
Alpha	176	208	222	42,838	52,746	58,847
Bravo	4	6	20	910	1,250	4,050
Charley	42	42	52	7997	8,450	10,520
Delta	102	115	141	20,001	23,194	28,346
Fox	84	82	60	16,200	16,500	12,000
Subtotal	408	453	495	87,946	102,140	111,763
Common use facilities						
Personnel - Education + Training @ 3.75 SF/P					1,699	1,856
Lunchroom + Lounge @ 4.50 SF/P					2,039	2,228
Subtotal common use facilities					3738	4,084
Total usable square feet required					105,878	115,847
X Rentable SF conversion factor (1.12 +/-)					118,583	129,749
(check your factor, about 1.10 to 1.15)						

* PRESENT footage is a total of space that would be required for present personnel, + special areas, + a circulation factor.

** MOVE-IN footage is based on a total of space requirements for projected personnel as of the move-in date, + minimal expansion, + special areas, + a circulation factor.

***FUTURE footage is based on total of space requirements for projected personnel (5) years after move-in, + special areas, + a circulation factor.

Figure 3.38

Furniture Inventory

Facility: _____ Location: _____

Department: _____ Floor: _____ Page: _____

Prepared By: _____ Phone: _____ Date: _____

Code	Qty	Description	Cond.	Width	Depth	Height	Return	Comments	Present Location	New Location

Codes

E1.0 Desks	E5.0 Executive seating	E 9.0 Filing Cabinets
E2.0 Credenzas	E6.0 Guest seating	E10.0 Storage/Shelving
E3.0 Tables	E7.0 Lounge furniture	E11.0 Miscellaneous
E4.0 Task seating	E8.0 Bookcases	

G. - Good   F. - Fair   P. - Poor

Figure 3.39

112

Example

E1.1 – 30 x 60 standard desk, black metal with plastic laminate top.
E1.2 – 30 x 66 standard desk, black metal with plastic laminate top.
E1.3 – 30 x 60 standard desk, black metal with wood top.

Follow the same procedure for all items, noting that any change in style, size, finish color, upholstery, or even minor details, such as hardware or bases, requires a new number within the series. Any variation should be assigned a separate item number. Even a 2" size difference in the field can be crucial. Once all the categories are established, count the total of each type and note it on the Inventory Form.

Step 3: Create a photo file. Once classifications are established, prepare a photo file of each major classification. This file should consist of either actual photos of the pieces, or catalog cut-sheets of the pieces when available.

Step 4: Organize for re-use. Now organize the furniture and assign each piece to the new plan. Group furniture pieces by style, manufacturer, color and size. (For example, all black metal desks could be grouped together.) Place items of similar design into groups, and use them in defined areas that will minimize the haphazard effect of mixing old and new furniture. Once the furniture assignment is complete, the code numbers should be noted on the plans. Note the new location on the inventory lists.

Step 5: Tag furniture. Prior to any move, temporarily tag all existing furniture so that movers can identify its future location. (Color coding is often a great help.) At the same time, examine all furniture needing repair or refinishing, and schedule both as necessary.

Step 6: Determine additional furniture needs. After all existing furniture has been assigned to the new facility, examine the space plans and identify all new furniture items that must be purchased to satisfy expanded requirements.

If it is necessary to refinish some existing furniture, there are two methods that can be used. The choice depends on the amount of time available. The better, and more costly, method of refinishing metal equipment is to have a

Example
E1.0 Desks
E2.0 Credenzas
E2.0 Tables
E4.0 Secretary Seating
E5.0 Swivel Seating
E6.0 Guest Chairs
E7.0 Lounge Furniture
E8.0 Bookcases
E9.0 Filing Cabinets
E10.0 Storage Shelving and Cabinets
E11.0 Miscellaneous
(In the example, an "E" code is used for existing furniture, while an "F" code is used for new furniture.)

Any variation in a classification means establishing a new number under that specific classification.

Figure 3.40

reputable furniture refinisher remove the items from the office and refinish them in his paint shop. This allows for the use of higher quality coatings which require special application conditions, but will produce a near factory-like finish.

The second method, for steel furniture only, is an electrostatic paint process, which can be done in the company's own offices after normal working hours. When refinished by this method, furniture can be used within 8 to 10 hours. Be sure to consult local building and fire codes before proceeding. Once again, electrostatic painting, while quick and able to be applied in the office, is not as durable as a shop-applied and baked-on finish.

Refinishing wood furniture used to be much easier and produce more satisfactory results than is likely to occur with a great deal of furniture manufactured today. This is a result of some of the manufacturing processes presently being employed. A great deal of furniture is constructed of sandwich type materials, consisting of an inner core which consistently used to be of solid wood, upon which was laminated a thin layer of wood veneer of the desired species. More recently this inner core may be of "particle board," consisting of ground particles of wood mixed with a glue-type binder and pressed into a solid board; or it may consist of paper laminates or honeycomb, forming the core, upon which has been laminated an extremely thin wood veneer — which may or may not be the wood species appearing in the finished product. The manufacturer may have applied to the veneer a ground coat of a filler-stain in the color of the desired finish. A wood grain is imprinted on the veneered surface, and the final finish applied. It is nearly impossible to determine how the surfaces were manufactured without beginning the stripping process. When a refinisher finds any of these conditions e.g., ultrathin veneers on a paper core or a printed grain, it is nearly impossible to produce a quality refinishing result. Ultrathin veneers on paper core will result in a very rapid absorption of the refinishing stains, resulting in a blotchy appearance. Many refinishers do not have capabilities in providing factory-printed grains, although most will be able to replicate some grains using hand applied techniques.

The manufacturing techniques described above are by no means limited to so-called cheap furniture. Some of the better furniture lines are manufactured using these processes.

Wood furniture touch-up can be performed on-site, either before or after relocation. There are balancing points of view as to which is the better time. If done before the move, there is less danger of damage to the surrounding areas, but greater likelihood of some damage due to the furniture itself due to the move. Moving the wood furniture first and performing a touch-up when it is at its new site seems better for the furniture, but places the new surroundings at risk. The best solution is to have all refinishing and as much touch-up done as is practical in a finishing shop.

Summary

The programming process may be one of the most important steps to this point. Carefully defining space standards and design elements begins to establish a cognizant policy of cost, image and space utilization. The process of seeking "needs and use" information through the requirements questionnaires and interviews helps to clarify the actual work process and requirements. The review and possible re-review of all of the program information will provide the project team with accurate and detailed information from which they may proceed with the design and specification process. It may also give management a new view on how work is performed within their company.

Chapter 4

Scheduling and Budgeting the Project

Introduction

This chapter deals with developing schedules and budgets, two of the most essential elements of a project. Several forms are provided to help accomplish both tasks as accurately as possible. They are explained here. The blank forms can be found in the Master Forms Section.

Of necessity, this material is complex. It may be necessary to review each section several times in order to fully understand all aspects of these two vital processes.

How To Begin

As soon as a facility change has been decided upon, appoint someone to act as the coordinator. This person will be responsible for:

- Assisting in coordinating the various aspects of the work flow;
- Coordinating the review and approval process;
- Assisting in establishing and maintaining the Project Schedule;
- Maintaining the Project Budget.

Creating a schedule with any real validity is a difficult and time-consuming process. Whether renovating one small department or moving a vary large facility from one site to another, the necessary work must be accomplished in a series of phases. Often, work on one phase cannot begin until work on an earlier phase is completed or has reached a specific stage of partial completion.

The forms are helpful for estimating how much time should be allowed for each phase, and in what order each should be performed.

At the point of final lease negotiations, whether for an addition or for completely new space, complete a preliminary schedule before agreeing on a projected occupancy date and before the lease is signed. The projected occupancy date may represent the time when the company must begin making rent payments. Once this date is fixed and the lease is signed, it is very unlikely that the date can be changed, even if actual move-in is delayed for any reason (unless it's the fault of the company's landlord and the lease provides for this eventuality). Therefore, it is necessary to determine this date as accurately as possible before the lease is signed.

If the company is interested in completing a change by a certain date, complete this scheduling process to see whether or not it will be feasible. If the company wishes to move or add on, but does not yet have any new space under lease, complete a preliminary project schedule to see by what date the final lease should be signed in order to achieve its move-in goal. If the project is simply a renovation, not an addition, complete a preliminary project schedule to determine when the renovations will be complete.

When it is time to begin the scheduling process, turn to the sample Project Schedule in Figures 4.1a and 4.1b. Have it available and refer to it while reading the explanatory text which appears on the following pages. Note that there are two different sheets. The Project Continuation Sheet should be attached to the first sheet. Additional Project Continuation Sheets can be added to the right-hand edges as needed for the project. Weeks are recommended as the best unit of time.

Gantt Charts and Critical Path Method (CPM) Networks

A computer-generated Gantt chart for a typical 50,000 SF project is shown in Figure 4.2. Whether you use a computerized scheduling system [either Gantt (time/task bars) or Critical Path Method (CPM) network], preliminary scheduling tends to be easier and quicker if done manually, as preliminary schedules must often be redone a number of times. These early schedules should contain notes as information becomes more specific. As we begin to switch over to a more detailed and specific schedule, data can be transferred to computer-generated Gantt charts. Our preference in computerized scheduling is slanted toward Gantt charts for reasons of simplicity and ease of comprehension by most people dealing with the schedules. We have also found the computer software to be simpler to manipulate.

CPM network charts are probably more accurate as they create a network of tasks usually projected in days (or even hours), with a critical path defined which can flag scheduling conflicts. However, CPM scheduling can be somewhat difficult and time-consuming. For very large and complex jobs, CPM may be most appropriate. There are experienced CPM scheduling consultants available for those who choose not to develop these skills in-house. Although there are people who may be able to construct and calculate a CPM network manually, most CPMs are produced by computer. If the opportunity affords itself, try both systems. Select what is comfortable and produces a satisfactory schedule.

Both systems normally have the capability of producing note or recap information related to each line item task. Once completed, these systems provide powerful project management tools and are surprisingly easy to monitor and update.

A manual format Gantt-type, or bar chart, schedule form (as shown in Figure 4.1) may be useful in a meeting, or when a computer is not available. Again, the original form sheets can be found in the Master Forms Section. Note that there are two different sheets involved. The Project Continuation Sheet should be attached to the first sheet. Additional Project Continuation Sheets can be added to the right-hand edges as needed for the project. Weeks are the recommended time unit, unless the project is of very short duration or you wish to monitor a critical phase, in which case switching to days might be warranted. Monthly charts are excellent for preliminary use to cover time periods counted in many months or years; but for day-to-day monitoring schedules, units scaled in weeks seem to work the best.

Each line item has pairs of dots to aid the extension of the bar or line. Use the upper set of dots for the initial or anticipated schedule. Use the lower set for revisions forward or back.

Vertical lines should be drawn down to the "Notes" sections. These will be helpful in identifying and documenting key events. Key events can also be identified with a letter referring to the note.

Just above the "Note" section is the portion of the form used for coordinating the Project Schedule to the Critical Dates Schedule. This portion of the Project Schedule should be completed before extending the time-lines.

The Project Phases, Defined

As evident in the samples, a Project Schedule is used to plan all necessary project phases when renovating, adding on, or moving to a new facility. In the text that follows, each phase is described, including an indication of when it begins, an explanation of how long it may take (in calendar weeks), a statement of the person or persons who should perform each of the necessary tasks, and a general idea of the level of the person or people who need to approve each step (write in the specific people in the company who will be responsible for approving each step). As the elements involved are complex, again, it is recommended that the text be read twice — first, to gain an overview of phase order and magnitude, then slowly, step-by-step. At each step, check the number that appears in the left-hand margin and locate the corresponding number on the sample schedule.

Note that the individual steps are listed phase-by-phase. If you have difficulty in locating a number, simply look back in the relevant text and find the nearest heading typed in capital letters above the step, then run your eye along the corresponding horizontal line on the sample. (For example, Step 14, *FINAL COST ACCOUNTING*, appears under *SCHEDULING AND BUDGETING*. Look along the *FINAL COST ACCOUNTING* horizontal line to find the number "14" near the left-hand margin.

As is necessary with any chart or schedule, review the legend to clarify the meaning of each symbol used.

Evaluating Space Requirements

1. **Estimating Square Footage Needed**
   When begins: Upon determination of a real need.
   Time required: Varies; consult interior designer.
   Who performs task: Coordinator, Facility Manager.
   Who approves: Officer in charge of facility.

2. **Estimating Preliminary Space Needs**
   (and orienting staff to the survey process).
   When begins: After market analysis is complete and preliminary lease negotiations have begun, but prior to the start of serious lease negotiations.
   Time required: Two to three days.
   Who performs task: Coordinator, Facility Manager.
   Who approves: Not necessary.

3. **Filling Out the Survey Questionnaires**
   When begins: After completion of orientation, but prior to conclusion of lease negotiations.
   Time required: 1 week.
   Who performs task: Each department.
   Who approves: Facility Manager.

4. **Reviewing and Tabulating Questionnaires**
   When begins: After all questionnaires are filled out, but prior to conclusion of lease negotiations.
   Time required: 2 weeks for first 50,000 SF; 1/2 week for each additional 50,000 SF.

Project Schedule

Facility: _____   Location: _____

Prepared By: _____   Phone: _____ Date: _____

Landlord: _____   Contact: _____

Architect: _____   Move-In Date: _____

No.	Task Description	Calendar Period
		Days, Weeks, Months (circle one)

Critical Date Numbers :

Notes :

Figure 4.1a

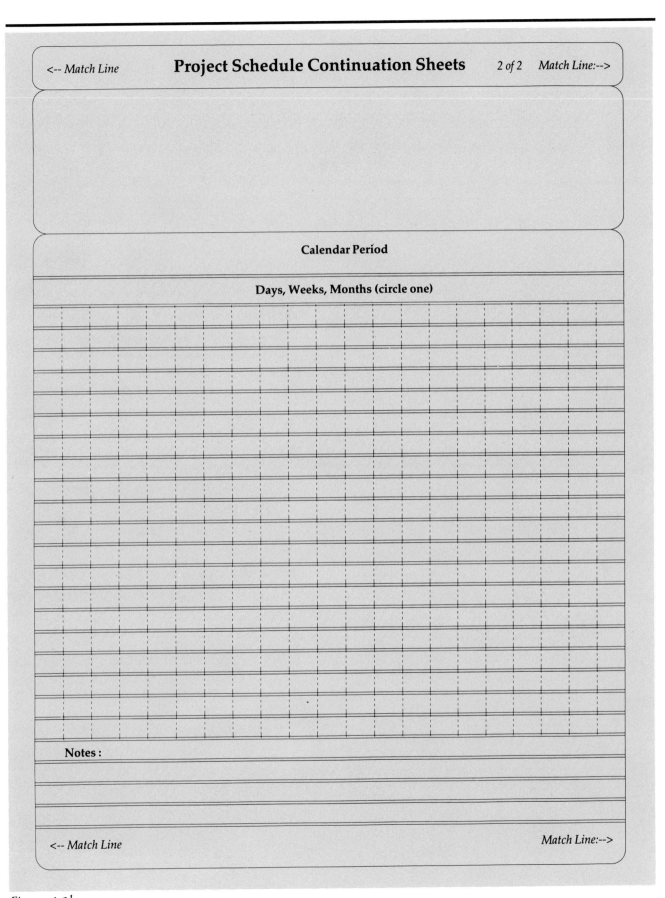

Figure 4.1b

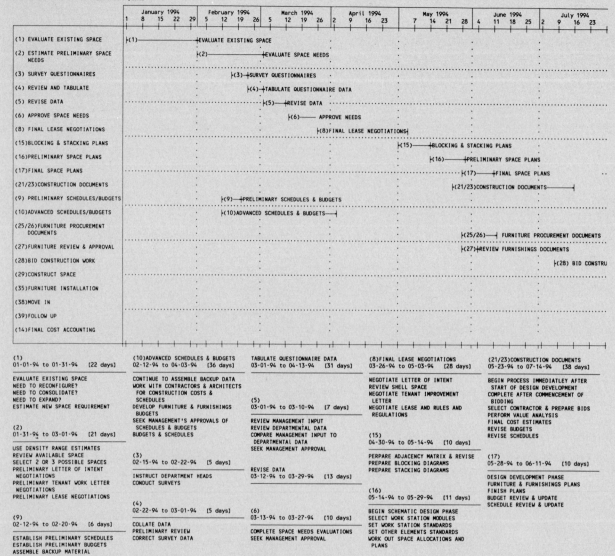

TYPICAL SCHEDULE - TO ESTABLISH CRITICAL DATES

ASSUMED: 50,000 SF PROJECT
NOTE:
FOR EACH ADDITIONAL 25,000 SF
ADD 3 TO 4 DAYS FOR DRAWINGS (YOUR ARCHITECT MAY BE ABLE TO SHORTEN)
ADD 2 WEEKS FOR CONSTRUCTION (THIS WILL BE FAIRLY CONSISTENT)
BEGINNING WITH START OF PRELIMINARY PLANS

	January 1994	February 1994	March 1994	April 1994	May 1994	June 1994	July 1994
(1) EVALUATE EXISTING SPACE	|(1)————————|EVALUATE EXISTING SPACE					
(2) ESTIMATE PRELIMINARY SPACE NEEDS		|(2)————————|EVALUATE SPACE NEEDS				
(3) SURVEY QUESTIONNAIRES			|(3)——|SURVEY QUESTIONNAIRES			
(4) REVIEW AND TABULATE			|(4)—|TABULATE QUESTIONNAIRE DATA			
(5) REVISE DATA			|(5)——| REVISE DATA			
(6) APPROVE SPACE NEEDS				(6)——— APPROVE NEEDS			
(8) FINAL LEASE NEGOTIATIONS				(8)	FINAL LEASE NEGOTIATIONS			
(15)BLOCKING & STACKING PLANS				|(15)——|BLOCKING & STACKING PLANS		
(16)PRELIMINARY SPACE PLANS				|(16)——|PRELIMINARY SPACE PLANS		
(17)FINAL SPACE PLANS					|(17)—|FINAL SPACE PLANS	
(21)(23)CONSTRUCTION DOCUMENTS					|(21/23)CONSTRUCTION DOCUMENTS————|	
(9) PRELIMINARY SCHEDULES/BUDGETS		|(9)—|PRELIMINARY SCHEDULES & BUDGETS				
(10)ADVANCED SCHEDULES/BUDGETS			(10)	ADVANCED SCHEDULES & BUDGETS——				
(25)(26)FURNITURE PROCUREMENT DOCUMENTS						(25/26)—— FURNITURE PROCUREMENT DOCUMENTS	
(27)FURNITURE REVIEW & APPROVAL					|(27)|REVIEW FURNISHINGS DOCUMENTS	
(28)BID CONSTRUCTION WORK							(28) BID CONSTRU
(29)CONSTRUCT SPACE							
(35)FURNITURE INSTALLATION							
(38)MOVE IN							
(39)FOLLOW UP							
(14)FINAL COST ACCOUNTING							

---

(1)
01-01-94 to 01-31-94   [22 days]

EVALUATE EXISTING SPACE
NEED TO RECONFIGURE?
NEED TO CONSOLIDATE?
NEED TO EXPAND?
ESTIMATE NEW SPACE REQUIREMENT

(2)
01-31-94 to 03-01-94   [21 days]

USE DENSITY RANGE ESTIMATES
REVIEW AVAILABLE SPACE
SELECT 2 OR 3 POSSIBLE SPACES
PRELIMINARY LETTER OF INTENT
   NEGOTIATIONS
PRELIMINARY TENANT WORK LETTER
   NEGOTIATIONS
PRELIMINARY LEASE NEGOTIATIONS

(9)
02-12-94 to 02-20-94   [6 days]

ESTABLISH PRELIMINARY SCHEDULES
ESTABLISH PRELIMINARY BUDGETS
ASSEMBLE BACKUP MATERIAL

(10)ADVANCED SCHEDULES & BUDGETS
02-12-94 to 04-03-94   [36 days]

CONTINUE TO ASSEMBLE BACKUP DATA
WORK WITH CONTRACTORS & ARCHITECTS
   FOR CONSTRUCTION COSTS &
   SCHEDULES
DEVELOP FURNITURE & FURNISHINGS
   BUDGETS
SEEK MANAGEMENT'S APPROVALS OF
   SCHEDULES & BUDGETS
BUDGETS & SCHEDULES

(3)
02-15-94 to 02-22-94   [5 days]

INSTRUCT DEPARTMENT HEADS
CONDUCT SURVEYS

(4)
02-22-94 to 03-01-94   [5 days]

COLLATE DATA
PRELIMINARY REVIEW
CORRECT SURVEY DATA

TABULATE QUESTIONNAIRE DATA
03-01-94 to 04-13-94   [31 days]

(5)
03-01-94 to 03-10-94   [7 days]

REVIEW MANAGEMENT INPUT
REVIEW DEPARTMENTAL DATA
COMPARE MANAGEMENT INPUT TO
   DEPARTMENTAL DATA
SEEK MANAGEMENT APPROVAL

REVISE DATA
03-12-94 to 03-29-94   [13 days]

(6)
03-13-94 to 03-27-94   [10 days]

COMPLETE SPACE NEEDS EVALUATIONS
SEEK MANAGEMENT APPROVAL

(8)FINAL LEASE NEGOTIATIONS
03-26-94 to 05-03-94   [28 days]

NEGOTIATE LETTER OF INTENT
REVIEW SHELL SPACE
NEGOTIATE TENANT IMPROVEMENT
   LETTER
NEGOTIATE LEASE AND RULES AND
   REGULATIONS

(15)
04-30-94 to 05-14-94   [10 days]

PERPARE ADJACENCY MATRIX & REVISE
PREPARE BLOCKING DIAGRAMS
PREPARE STACKING DIAGRAMS

(16)
05-14-94 to 05-29-94   [11 days]

BEGIN SCHEMATIC DESIGN PHASE
SELECT WORK STATION MODULES
SET WORK STATION STANDARDS
SET OTHER ELEMENTS STANDARDS
WORK OUT SPACE ALLOCATIONS AND
   PLANS

(21)(23)CONSTRUCTION DOCUMENTS
05-23-94 to 07-14-94   [38 days]

BEGIN PROCESS IMMEDIATLEY AFTER
   START OF DESIGN DEVELOPMENT
COMPLETE AFTER COMMENCEMENT OF
   BIDDING
SELECT CONTRACTOR & PREPARE BIDS
PERFORM VALUE ANALYSIS
FINAL COST ESTIMATES
REVISE BUDGETS
REVISE SCHEDULES

(17)
05-28-94 to 06-11-94   [10 days]

DESIGN DEVELOPMENT PHASE
FURNITURE & FURNISHINGS PLANS
FINISH PLANS
BUDGET REVIEW & UPDATE
SCHEDULE REVIEW & UPDATE

Figure 4.2

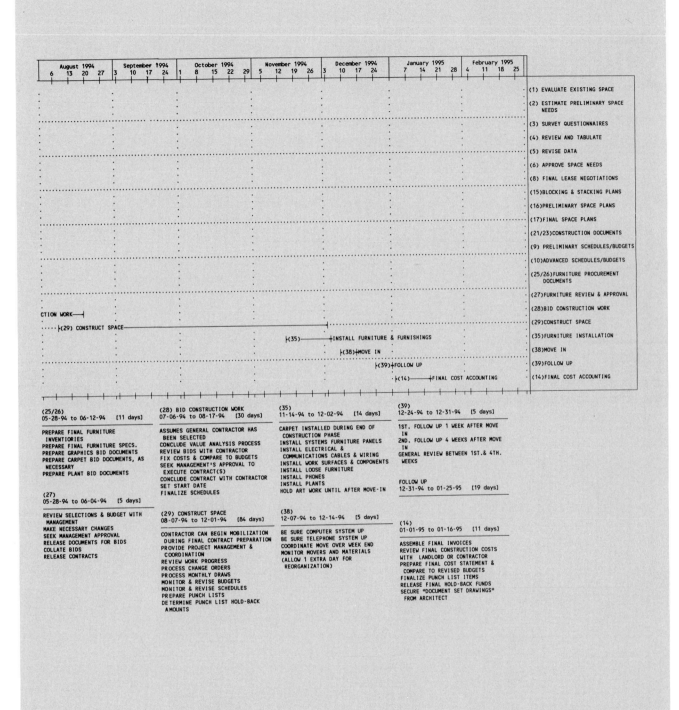

	August 1994			September 1994			October 1994				November 1994			December 1994			January 1995			February 1995			
6	13	20	27	3	10	17	24	1	8	15	22	29	5	12	19	26	3	10	17	24	7	14	21	28	4	11	18	25

(1) EVALUATE EXISTING SPACE

(2) ESTIMATE PRELIMINARY SPACE NEEDS

(3) SURVEY QUESTIONNAIRES

(4) REVIEW AND TABULATE

(5) REVISE DATA

(6) APPROVE SPACE NEEDS

(8) FINAL LEASE NEGOTIATIONS

(15)BLOCKING & STACKING PLANS

(16)PRELIMINARY SPACE PLANS

(17)FINAL SPACE PLANS

(21/23)CONSTRUCTION DOCUMENTS

(9) PRELIMINARY SCHEDULES/BUDGETS

(10)ADVANCED SCHEDULES/BUDGETS

(25/26)FURNITURE PROCUREMENT DOCUMENTS

(27)FURNITURE REVIEW & APPROVAL

(28)BID CONSTRUCTION WORK

(29)CONSTRUCT SPACE

(35)FURNITURE INSTALLATION

(38)MOVE IN

(39)FOLLOW UP

(14)FINAL COST ACCOUNTING

CTION WORK

(29) CONSTRUCT SPACE

(35) INSTALL FURNITURE & FURNISHINGS

(38) MOVE IN

(39) FOLLOW UP

(14) FINAL COST ACCOUNTING

(25/26)
05-28-94 to 06-12-94   [11 days]

PREPARE FINAL FURNITURE
  INVENTORIES
PREPARE FINAL FURNITURE SPECS.
PREPARE GRAPHICS BID DOCUMENTS
PREPARE CARPET BID DOCUMENTS, AS
  NECESSARY
PREPARE PLANT BID DOCUMENTS

(27)
05-28-94 to 06-04-94   [5 days]

REVIEW SELECTIONS & BUDGET WITH
  MANAGEMENT
MAKE NECESSARY CHANGES
SEEK MANAGEMENT APPROVAL
RELEASE DOCUMENTS FOR BIDS
COLLATE BIDS
RELEASE CONTRACTS

(28) BID CONSTRUCTION WORK
07-06-94 to 08-17-94   [30 days]

ASSUMES GENERAL CONTRACTOR HAS
  BEEN SELECTED
CONCLUDE VALUE ANALYSIS PROCESS
REVIEW BIDS WITH CONTRACTOR
FIX COSTS & COMPARE TO BUDGETS
SEEK MANAGEMENT'S APPROVAL TO
  EXECUTE CONTRACT(S)
CONCLUDE CONTRACT WITH CONTRACTOR
SET START DATE
FINALIZE SCHEDULES

(29) CONSTRUCT SPACE
08-07-94 to 12-01-94   [84 days]

CONTRACTOR CAN BEGIN MOBILIZATION
  DURING FINAL CONTRACT PREPARATION
PROVIDE PROJECT MANAGEMENT &
  COORDINATION
REVIEW WORK PROGRESS
PROCESS CHANGE ORDERS
PROCESS MONTHLY DRAWS
MONITOR & REVISE BUDGETS
MONITOR & REVISE SCHEDULES
PREPARE PUNCH LISTS
DETERMINE PUNCH LIST HOLD-BACK
  AMOUNTS

(35)
11-14-94 to 12-02-94   [14 days]

CARPET INSTALLED DURING END OF
  CONSTRUCTION PHASE
INSTALL SYSTEMS FURNITURE PANELS
INSTALL ELECTRICAL &
  COMMUNICATIONS CABLES & WIRING
INSTALL WORK SURFACES & COMPONENTS
INSTALL LOOSE FURNITURE
INSTALL PHONES
INSTALL PLANTS
HOLD ART WORK UNTIL AFTER MOVE-IN

(38)
12-07-94 to 12-14-94   [5 days]

BE SURE COMPUTER SYSTEM UP
BE SURE TELEPHONE SYSTEM UP
COORDINATE MOVE OVER WEEK END
MONITOR MOVERS AND MATERIALS
  (ALLOW 1 EXTRA DAY FOR
  REORGANIZATION)

(39)
12-24-94 to 12-31-94   [5 days]

1ST. FOLLOW UP 1 WEEK AFTER MOVE
  IN
2ND. FOLLOW UP 4 WEEKS AFTER MOVE
  IN
GENERAL REVIEW BETWEEN 1ST.& 4TH.
  WEEKS

FOLLOW UP
12-31-94 to 01-25-95   [19 days]

(14)
01-01-95 to 01-16-95   [11 days]

ASSEMBLE FINAL INVOICES
REVIEW FINAL CONSTRUCTION COSTS
  WITH LANDLORD OR CONTRACTOR
PREPARE FINAL COST STATEMENT &
  COMPARE TO REVISED BUDGETS
FINALIZE PUNCH LIST ITEMS
RELEASE FINAL HOLD-BACK FUNDS
SECURE "DOCUMENT SET DRAWINGS"
  FROM ARCHITECT

Figure 4.2 (*continued*)

Who performs task: Professional staff or consultants as directed by Coordinator.

Who approves: Facility Manager.

5. **Revising Completed Survey Tabulations**

When begins: Approximately 1-1/2 weeks after completion of tabulations.

Time required: 1/2 week.

Who performs task: Coordinator, Facility Manager.

Who approves: Officer in charge of facility.

6. **Final Review and Revisions of Space Needs**

When begins: After substantial completion of lease negotiations, but just prior to the formal signing of lease.

Time required: 2 weeks for first 50,000 SF; 1/2 week for each additional 20,000 SF.

Who performs task: Coordinator, Facility Manager.

Who approves: Officer in charge of facility.

Note: *This is an essential review. Do not omit!*

Selecting Leasing Space

7. **Preliminary Market Study**

When begins: After estimating square footage needed.

Time required: Varies; consult designer or broker.

Who performs task: Coordinator, professional staff or consultants, Facility Manager.

Who approves: Officer in charge of facility.

8. **Final Lease Negotiations**

When begins: After review/approval of Space Needs Questionnaires.

Time required: Varies; consult interior designer.

Who performs task: Professional staff or consultants, directed by Coordinator, with advice of an attorney.

Who approves: Officer in charge of facility.

Note: *"Lease Negotiations Substantially Completed," and "Lease Signed" points occur here.*

Scheduling and Budgeting

9. **Preliminary Schedule and Budget Estimates**

When begins: After completion of square footage estimate, but prior to any other phase.

Time required: Varies; consult interior designer.

Who performs task: Professional staff or consultants, with direction from Coordinator, Facility Manager.

Who approves: Officer in charge of facility.

10. **Developing Project Schedules and Budgets**

When begins: After substantial completion of lease documents, but just prior to signing the final lease.

Time required: Varies; consult designer.

Who performs task: Professional staff or consultants, with information provided by Coordinator, Facility Manager.

Who approves: Officer in charge of facility.

Schedule and Budget Revisions

11. **Planning Phase Review**

When begins: After substantial completion of the planning process.

Time required: Varies; consult interior designer.

Who performs task: Coordinator, Facility Manager.

Who approves: Officer in charge of facility.

12. **Design Phase Review**
    When begins: After substantial completion of design process.
    Time required: Varies; consult interior designer.
    Who performs task: Coordinator, Facility Manager.
    Who approves: Officer in charge of facility.

13. **Construction Phase Review**
    When begins: At conclusion of bidding the job, prior to the award of construction contracts.
    Time required: Varies; consult interior designer.
    Who performs task: Professional staff or consultants, with review by Coordinator, Facility Manager.
    Who approves: Officer in charge of facility.

14. **Final Cost Accounting**
    When begins: Approximately two to three weeks after move-in to the new facility is complete, with all items delivered and installed.
    Time required: Varies; consult interior designer.
    Who performs task: Coordinator, Facility Manager.
    Who approves: Officer in charge of facility.

Planning the Office Space

15. **Blocking and Stacking Diagrams**
    When begins: After space requirements are fully evaluated and lease is signed.
    Time required: 2 weeks for the first 90,000 SF; 1/2 week for each additional 20,000 SF.
    Who performs task: Professional staff or consultants with data provided by Coordinator, Facility Manager.
    Who approves: Officer in charge of facility.

    Note: If time pressures are severe, this process can begin after substantial completion of lease negotiations, but prior to signing.

Space Planning

16. **Preliminary Space Plans**
    When begins: After space requirements are fully evaluated; negotiations for any additional new space are substantially completed; and blocking and stacking diagrams are approved.
    Time required: 2 weeks for the first 50,000 SF; 1 week for each additional 25,000 SF.
    Who performs task: Professional staff or consultants as directed by Coordinator and Facility Manager.
    Who approves: Officer in charge of facility.

    Note: No space planning should be started until the lease documents are formally signed. If the lease has not been formally signed by the time the blocking and stacking diagrams are complete, the move-in may be delayed. Extend all remaining phases of the project by a number of days equal to the number of days between completion of blocking and stacking diagrams and signing of the lease.

17. **Final Space Plans**
    When begins: After review and approval of preliminary space plans.
    Time required: 1 week for first 50,000 SF; 1/2 week for each additional 25,000 SF.
    Who performs task: Professional staff or consultants as directed by Coordinator and Facility Manager.
    Who approves: Officer in charge of facility.

18. **Revised Space Plans (if necessary)**
When begins: After review and approval of final space plans. Time required: 1 week for first 50,000 SF; 1/2 week for each additional 25,000 SF.
Who performs task: Professional staff or consultants as directed by Coordinator and Facility Manager.
Who approves: Officer in charge of facility.

Designing the Office Space

19. **Design**
When begins: After review and approval of the substantially completed space plan.
Time required: 2 weeks for first 50,000 SF; 1 week for each additional 25,000 SF.
Who performs task: Professional staff or consultants as directed by Coordinator and Facility Manager.
Who approves: Officer in charge of facility.

20. **Review and Approval**
When begins: At the conclusion of the design process.
Time required: 1/2 week for the first 50,000 SF; 1/2 week for each additional 50,000 SF.
Who performs task: Professional staff or consultants as directed by Coordinator and Facility Manager.
Who approves: Officer in charge of facility.

Note: This final review is in addition to any weekly/monthly reviews which may also be required. After review and approval of the office design, but prior to the start of the construction documents, the plans should be reviewed, if possible, by all regulatory authorities having jurisdiction over the project. At the end of this phase, verify with all major manufacturers the extent of lead times required for manufacturing and shipping. Failure to allow adequate lead times can invalidate the occupancy date. If long lead times may delay planned furniture installation, letters of intent should be issued now, in order to guarantee manufacturing time slots and delivery dates.

Construction Documents

21. **Background Sheet Preparation**
When begins: After lease negotiations are substantially completed, but prior to the beginning of space planning.
Time required: 1/2 week for the first 50,000 SF; varies for each additional 25,000 SF; consult designer.
Who performs task: Professional staff or consultants as directed by Coordinator and Facility Manager.
Who approves: Officer in charge of facility.

22. **Production**
When begins: After review and approval of the substantially completed office design.
Time required: 4 weeks for the first 50,000 SF; 2 weeks for each additional 25,000 SF.
Who performs task: Professional staff or consultants as directed by Coordinator and Facility Manager.
Who approves: Officer in charge of facility.

23. **Coordination and Clarification**
When begins: Upon substantial completion of the construction documents.
Time required: 1 week, plus.
Who performs task: Professional staff or consultants as directed by Coordinator and Facility Manager.
Who approves: Officer in charge of facility.

Note: During this time, all documents are thoroughly reviewed for such things as completeness and correct cross-referencing for details. At the conclusion of this phase, the construction documents should be totally complete and ready for review.

24. **Review and Approval**

When begins: After completion of the construction documents, but prior to the release for bid.

Time required: Consult designer, about 1 week.

Who performs task: Professional staff or consultants as directed by Coordinator and Facility Manager.

Who approves: Officer in charge of facility.

Note: The documents will be thoroughly reviewed by the Coordinator. This final review is in addition to the weekly/monthly reviews which may also be required. The Building landlord or Building Architect may also require time for review and approval before these documents may be released for bid.

Furniture Procurement Documents and Orders

25. **Production**

When begins: 2 weeks after the start of construction documents, but prior to any purchasing.

Time required: 3 weeks for first 50,000 SF; 1 week for each additional 25,000 SF.

Who performs task: Professional staff or consultants as directed by Coordinator, Facility Manager.

Who approves: Officer in charge of facility.

26. **Review and Approval**

When begins: After the procurement documents are complete, but prior to any purchasing.

Time required: 1 week.

Who performs task: Coordinator, Facility Manager.

Who approves: Officer in charge of facility.

27. **Addenda**

When begins: As required after review/approval of procurement documents, but prior to any purchasing.

Time required: 1 week, only if necessary.

Who performs task: Coordinator, Facility Manager.

Who approves: Officer in charge of facility.

Note: The "Furniture Order Final" stage occurs here.

Construction

28. **Bidding the Construction Job**

When begins: After review/approval of the documents by Coordinator.

Time required: 2 to 6 weeks.

Who performs task: Coordinator.

Who approves: Officer in charge of facility.

Note: Bidding may take 2 to 6 weeks, depending on the landlord/contractor or client/contractor relationship, and the job size and complexity. Included in the estimate is an allowance of 1 week for reviewing bids and awarding the construction contract.

29. **Construction of Space**

When begins: After the contract is awarded.

Time required: 12 weeks for first 50,000 SF; 1 to 2 weeks for each additional 25,000 SF.

Who performs task: General Contractor under direction of landlord or Coordinator/Facilities Manager.

Who approves: Coordinator.

Note: When union-recognized holidays occur during the construction period, extend the estimated construction time. Consult designer on how to do this.

Signage & Art

30. Design and Procurement Documents

When begins: After review/approval of office design.
Time required: 2 weeks for the first 50,000 SF; 1/2 week for each additional 50,000 SF.
Who performs task: Professional staff or consultants as directed by Coordinator.
Who approves: interior signage – Facility Manager, exterior signage – Coordinator, art – officer in charge of facility.

31. Manufacturing Signage

When begins: After review/approval of the proposed graphic program.
Time required: 6 weeks for the first 50,000 SF; 1 week for each additional 20,000 SF.
Who performs task: Manufacturers, monitored by Coordinator.
Who approves: Not necessary.

Note: Verify manufacturing lead time required for order date.

32. Installation of Signage

When begins: After delivery of manufactured items and after substantial completion of furniture installation.
Time required: 1 week for the first 50,000 SF; 1/2 week for each additional 40,000 SF.
Who performs task: Manufacturers, directed by Coordinator.
Who approves: Coordinator.

Plants & Containers

33. Design and Procurement Documents

When begins: After review and approval of office design, but prior to conclusion of construction documents.
Time required: 2 weeks for the first 50,000 SF; 1/2 week for each additional 20,000 SF.
Who performs task: Professional staff or consultants as directed by Coordinator, Facility Manager.
Who approves: Officer in charge of facility.

Note: Design and procurement documents may be scheduled, if necessary, any time after review/approval of the office design. The only requirement is that this phase be completed to allow adequate time for review/approval, bid, order time and installation. Consult Interior Designer or Coordinator/Facilities Manager for further information.

34. Review/Approval, Bid, Order/Delivery

When begins: After completion of design and procurement documents, but prior to installation.
Time required: 2 weeks for the first 50,000 SF; 1/2 week for each additional 50,000 SF.
Who performs task: Coordinator/Facilities Manager.
Who approves: Officer in charge of facility.

Note: determine order/delivery time by contacting a local plant service or distributor.

35. Installation

When begins: After installation of furniture is substantially complete, but prior to move-in.
Time required: 2 weeks for the first 50,000 SF; 1/2 week for each additional 50,000 SF.

Who performs task: Vendor.

Who approves: Coordinator.

Note: Verify time required with local plant service or distributor.

Furniture Fabrication

36. When begins: After initial order has been placed with manufacturer, and deposit (if required) has been paid.

    Time required: Verify with manufacturer as early as possible. Consult designer for schedule information.

    Who performs task: Furniture manufacturers monitored by Coordinator.

    Who approves: Not necessary.

    Note: Time shown on schedule should allow for lead time and shipping time, as well as actual manufacturing time.

Installation and Moving In

37. **Installation**

    When begins: After completion of major interior construction (including complete carpet installation).

    Time required: 2 weeks for the first 50,000 SF; 1/2 week for each additional 15,000 SF.

    Who performs task: Carpeting or general contractor, furniture dealers or installers.

    Who approves: Coordinator.

38. **Moving In**

    When begins: After completed installation.

    Time required: 3 days for first 50,000 SF; (Friday and weekend) 1 day for each additional 20,000 SF.

    Who performs task: Moving company.

    Who approves: Coordinator.

    Note: This estimate is based on a local move. Longer distances require consultation with your mover and designers for appropriate time allocations. The "Facility Operational" stage occurs here.

39. **Following up After Occupancy**

    When begins: At one week, one month, and six months after move-in.

    Time required: Varies.

    Who performs task: Coordinator.

    Who approves: Officer in charge of facility.

    When preparing a schedule, begin by inserting appropriate calendar dates above each vertical line. This makes it possible to identify the point in calendar time when each step should take place.

Critical Dates Form

Once the Project Schedule is complete, you will be in a position to complete what may well be the single most useful form connected with the project. It is called *Critical Dates*, and appears in Figure 4.3. (For a copy, refer to Master Forms Section.) Verbalizing the schedule forces the project manager to think through each portion of the project and to then record dates for each segment. The completed Critical Dates schedule should be distributed to any party involved in the timing of the project. Each person should be asked to voice any objections or to raise questions upon receipt of the schedule. It is better that concerns be raised at the outset rather than when the job is under way and there are possibilities for delay.

Before attempting to fill out a Critical Dates form, be sure the Project Schedule has been fully developed, item by item. The Critical Dates form cannot be completed without it.

Critical Dates

Facility: _____

Location: _____

Prepared By: _____

Phone: _____ Date: _____

Landlord: _____

Contact: _____

Architect: _____

Move-In Date: _____

DATES :

TASKS :

Estimated ___ Revised _____ Actual _____	1. Review/approve completed space requirements survey
Estimated ___ Revised _____ Actual _____	2. Conclude market survey; begin lease negotiations
Estimated ___ Revised _____ Actual _____	3. Review/approve final space requirements survey totals
Estimated ___ Revised _____ Actual _____	4. Substantially complete lease negotiations
Estimated ___ Revised _____ Actual _____	5. Review/approve project schedule
Estimated ___ Revised _____ Actual _____	6. Review/approve preliminary estimated budget
Estimated ___ Revised _____ Actual _____	7. Complete lease negotiations and execute lease
Estimated ___ Revised _____ Actual _____	8. Review/approve adjacency matrices
Estimated ___ Revised _____ Actual _____	9. Review/approve stacking and blocking plans
Estimated ___ Revised _____ Actual _____	10. Complete preliminary space plans
Estimated ___ Revised _____ Actual _____	11. Review/approve final space plans
Estimated ___ Revised _____ Actual _____	12. Obtain preliminary building department approval of space plans
Estimated ___ Revised _____ Actual _____	13. Select phone system
Estimated ___ Revised _____ Actual _____	14. Complete design of space
Estimated ___ Revised _____ Actual _____	15. Review/approve final design of space
Estimated ___ Revised _____ Actual _____	16. Confirm delivery dates, issue letters of intent (if necessary)
Estimated ___ Revised _____ Actual _____	17. Review/approve signage and graphics

Figure 4.3

DATES :

TASKS :

Estimated ___ Revised _____ Actual _____   18. Review/approve interior landscaping

Estimated ___ Revised _____ Actual _____   19. Review/approve art program

Estimated ___ Revised _____ Actual _____   20. Review/approve final budget

Estimated ___ Revised _____ Actual _____   21. Order furniture and confirm delivery date

Estimated ___ Revised _____ Actual _____   22. Select mover

Estimated ___ Revised _____ Actual _____   23. Order carpet

Estimated ___ Revised _____ Actual _____   24. Review/approve construction documents

Estimated ___ Revised _____ Actual _____   25. Release construction documents for bid

Estimated ___ Revised _____ Actual _____   26. Review/approve bid; select contractors; award contracts

Estimated ___ Revised _____ Actual _____   27. Construction substantially complete

Estimated ___ Revised _____ Actual _____   28. Complete floor covering installation

Estimated ___ Revised _____ Actual _____   29. Bring computer room up and cut over

Estimated ___ Revised _____ Actual _____   30. Begin furniture installation

Estimated ___ Revised _____ Actual _____   31. Begin phone instrument installation

Estimated ___ Revised _____ Actual _____   32. Begin installation of data wiring

Estimated ___ Revised _____ Actual _____   33. Complete housecleaning; begin packing

Estimated ___ Revised _____ Actual _____   34. Begin move

Estimated ___ Revised _____ Actual _____   35. Facility operational

Figure 4.3 (*continued*)

The Critical Dates form summarizes all vital scheduling information in an easily understood format. Begin by filling out the name of the facility and its location. Add general information such as names, addresses, and phone numbers of people involved in the project as each becomes known. The critical dates themselves may be determined by referring to the Project Schedule. They are listed and explained below:

1. Review/approve completed Space Requirements Survey.
   Date: After completion of Survey, but prior to beginning of lease negotiations.

2. Conclude market survey; begin lease negotiations.
   Date: After completion of Space Requirements Survey, but prior to start of negotiations.

3. Review/approve Space Requirements Survey totals.
   Date: After completion of survey revisions, but prior to substantial completion of lease negotiations.

4. Substantially complete lease negotiations.
   Date: After the start of serious lease negotiations, but prior to the conclusion of lease negotiations.

5. Review/approve Project Schedule.
   Date: After Project Schedule has been formulated, but prior to the signing of the lease.

6. Review/approve Preliminary Budget Estimate.
   Date: After estimated budget has been defined, but prior to signing the lease.

7. Complete lease negotiations and execute lease.
   Date: After the conclusion of lease negotiations, and prior to all other phases.

8. Review/approve Adjacency Matrix.
   Date: After final review/approval of questionnaire, but just prior to blocking and stacking plans.

9. Review/approve blocking and stacking plans.
   Date: After completion of blocking and stacking plans, but prior to the beginning of space planning.

10. Complete preliminary space plan.
    Date: After lease is signed.

11. Review/approve final space plan.
    Date: After the completion of the final space plan, but prior to the start of design.

12. Obtain preliminary building department approval of space plans.
    Date: After review/approval of final space plan, but prior to the start of design.
    Note: This may not be possible in the company's jurisdiction, as building departments no longer have the staff for preliminary reviews, or the building department may review but not "approve."

13. Select communications/phone system.
    Date: After approval of space plan and preliminary plan check, but prior to preparation of construction documents (verify lead time with phone suppliers).

14. Complete design of space.
    Date: After all space plans have been reviewed and approved.

15. Review/approve final design of space.
    Date: After completion of final design.

16. Confirm delivery dates; issue letters of intent (if necessary).
    Date: After review/approval of design phase.

17. Review/approve signage and graphics.
    Date: After completion of signage/graphics design and procurement documents.

18. Review/approve interior landscaping (and service contracts, if necessary).
    Date: After completion of plant landscaping design and procurement documents.
19. Review/approve art program.
    Date: After completion of art program selections.
20. Review/approve final budget.
    Date: After all previous phases have been completed and budget information analyzed.
21. Order furniture and confirm delivery date.
    Date: After approval of design, procurement documents, and furniture budget.
22. Select mover.
    Date: After review/approval of design phase. (Verify lead time required with Coordinator or officer in charge of facility.)
23. Order carpet.
    Date: After review/approval of design phase. (Verify lead time required with Coordinator or officer in charge of facility.)
24. Review/approve construction documents.
    Date: After construction documents are completed and coordinated.
25. Release construction documents for bid.
    Date: After approval for release by Coordinator.
26. Review/approve bid; select contractors, award contracts.
    Date: After receiving bid and checking for completeness.
27. Construction Substantially Complete.
28. Complete carpet installation.
    Note: Some contractors like to separate carpet purchasing from installation. This can prevent potential labor problems, but without a carpet vendor representative on the job, this procedure can lead to disputes over installation techniques, damaged or soiled carpeting.
29. Complete computer room and start bringing systems up for switch-over from old facility.
30. Begin furniture installation.
    Date: After substantial completion of carpet installation.
31. Begin communications/phone instrument installation.
    Date: After substantial completion of furniture installation.
32. Begin installation of data wiring.
33. Complete housecleaning; begin packing.
34. Begin move.
    Date: After completion of packing; after movers have tagged all items.
35. Facility operational.
    Date: After completion of moving.
    Note: Critical dates do not necessarily occur in numerical order. The sequence in which critical dates occur depends on each individual project. This is because critical dates are often dependent on lead times necessary to accomplish various project steps.

When entering the actual, critical dates for the project on the Project Schedule as well as on the Critical Dates form, it is easier to enter them above the appropriate time symbol (asterisk or bar), rather than at the bottom of the schedule.

The main reason for the existence of the Critical Dates form is that it provides essential checkpoints for the project. If one aspect of the necessary work begins to slip behind schedule, that slippage must be eliminated so that the next critical date can be met on schedule. If the slippage cannot be eliminated, the probabilities are high that the whole schedule will slip, resulting in delayed occupancy and higher costs.

When someone asks that one of these dates be adjusted (either earlier or later), remember that changing one may have many subtle effects on the entire schedule. Most dates are interrelated.

Keep in mind the importance of making a permanent record on the form of the *date of its original issue* and of *each subsequent revision*. For example, if a strike occurred in the construction industry, it could delay the interior construction and thus, perhaps the entire project. If it is clear when looking at the form when it was last revised, you can be sure recent causes of delay have been taken into account.

It is also important to keep a careful list of the persons to whom copies of this form are sent: suppliers, vendors, artisans, etc. That way when dates must be changed (for example, in the case of a strike), a revised form can be sent to everyone on the list. If they must adjust their schedules accordingly, be sure to tell them so in writing at the same time they are sent the revised form.

Refining the Budget and Establishing Cost Control Methods

At the same time the Project Schedule is created, a detailed budget should also be prepared. This detailed budget will make it possible to evaluate the impact of planned changes on project costs. Once a final budget is established and approved by all the necessary parties, it becomes the control point for the project and should not be exceeded. Therefore, take particular care to thoroughly investigate all potential costs associated with the project. Do not proceed with further planning until the budget has been fully developed and approved.

Preliminary Budget Estimate
The first form, the Preliminary Budget Estimate, should be started as soon as the required number of square feet for the facility has been estimated. A sample Preliminary Budget Estimate form is reproduced in reduced size in Figure 4.4. A full-size version can be found in the Master Forms Section.

To begin preparing the Preliminary Budget Estimate, duplicate the Master Form. Then fill in each of the columns as appropriate after thoroughly reviewing the detailed explanations of the line item categories, shown on the form. The categories extending horizontally across the top of the preliminary budget estimate form should be filled in as follows:

Estimate Source:
Indicate the source of the estimated budget data shown in the next column. (Possible sources for budget estimates will be given under the description of each budget category.)

Estimated Budget:
This column will include cost estimates for each budget category specified in the left-hand column.

$/SF:
Indicate the dollar costs per usable square foot of each amount listed in the Estimated Budget column. The budget categories listed vertically on the left-hand side of the budget form under *Description* should be used as follows:

The categories extending vertically describe the various costs to be incurred: work to be done, services to be performed, materials to be purchased or project-related expenses.

General Construction:
This category covers all items which are in excess of building standards and are not covered by tenant allowances, including: additional quantities of wall partitioning, doors, frames, hardware, lighting, electrical and telephone equipment, acoustical ceiling, special millwork, floor covering, window covering, heating, ventilating and air-conditioning, plumbing and fire sprinkler, painting, and wall covering. In addition, this category should include items which are not normally provided by the landlord, such as special decorative lighting or cabinetry, inter-floor stairs, dumbwaiters, special

Preliminary Budget Estimate

Facility: _____ Location: _____

Prepared By: _____ Phone: _____ Date: _____

Owner: _____ Contact: _____

	Estimated	Actual	
Area			
Rentable			
Usable			
Description	**Estimate Source**	**Estimated Budget**	**$/SF**
General Construction			
Furniture and Fixtures			
Accessories			
Signage and Graphics			
Architectural			
Outside Consultants			
Miscellaneous and Unallocated Costs			
Telephone, Data and Communications			
Packing and Moving			
New Equipment and Installation			
Travel			
Legal			
Contingency			
Totals			

Figure 4.4

decorative wall coverings, glass partitions, plumbing, special air-conditioning equipment for the computer system, exhaust fans for conference rooms, and the general contractor's general conditions, overhead and profit.

Note: Possible sources for general construction estimates include local general contractors and building owners from whom proposals are being solicited.

Furniture and Fixtures:

Include the cost of any furniture the company plans to purchase in this category, such as: workstation systems furniture and components, conference room furniture, executive furniture, reception desks, chairs, and new freestanding file cabinets. In addition, include the cost of refinishing, re-upholstering, and repairing existing equipment to be relocated.

Note: For estimating purposes, the company's architect or space planner can often provide a reasonably accurate budget expressed in dollars per square foot.

Accessories:

Include such items as wastebaskets, ashtrays, letter trays, calendars, clocks, and the like.

Note: To make this estimate, go to a local office supply house or ask the company's space planner. Ask them to provide estimates based on the guidelines in the Accessories section.

Artwork:

Use this budget item to cover all artwork (i.e., paintings, sculpture, prints, tapestries, etc.) to be placed in reception areas, open work areas, private offices, conference rooms, and any other appropriate areas.

Note: For estimating purposes, an art consultant can, and probably should, provide an approximation. See section on Art.

Plants and Containers:

If the company will be renting plants for its new facility, include only the container or planter costs and installation charges. If plants will be purchased, include these charges as well. (If plants will be rented, be sure to add the rental charges to the monthly occupancy costs projection.) Note: For estimating purposes, visit two or three local indoor plant nurseries. See Chapter 14 on Plants and Containers.

Signage and Graphics:

Use this budget item to cover safety graphics, name plaques, all floor signs, directories, etc.

Note: For estimating purposes, the company's architect or space planner can provide approximations. See Chapter 13 on Signage and Graphics.

Exterior Signage:

This item covers the cost of placing a sign identifying the company or facility in one or more exterior locations.

Note: For estimating purposes, the company's architect or space planner can provide an approximation.

Outside Consultants:

The content of this category varies widely, depending on the method selected to accomplish the various tasks associated with the preparation of the new facility. Depending on the facility's size and complexity, this list may include: architects, designers, space planners, mechanical engineers, acoustical engineers, lighting and electrical engineers, computer specialists, attorneys, and consultants in the areas of furniture selection, communications, food service, plants, art, physical conditioning, medical facilities, construction, and moving. In their area of expertise, each may save the project considerable time and money, as well as helping to provide a superior product.

It is not necessary to retain every type of consultant possible. However, before making a final decision, find out if any of these consultants provide their services as part of their marketing efforts. Furniture, plants, food service, physical conditioning, communications and art are a few areas where consultant services are often packaged as a part of their services, if you purchase their products. These consultants are frequently as well qualified as any independent professional, although they typically will not be knowledgeable about building codes. It should be obvious that these product- or service-related consultants are going to be biased in favor of their particular products. Nonetheless, if it is determined that the products or services are essentially equal, the gratis consulting service can be most effective.

It is crucial that all the efforts of the company's consultants, however contracted (whether through the company or its architect), be *coordinated* through the company's architect or designer. This coordination is one of the key services performed by the architect. While it may be customary for the company to work with a particular computer installation consultant, with whom it has had a relationship for many years, it is still very important that the architect *coordinate* all work to be performed by this, and any other consultant to avoid overlooked information and potential conflicts.

A major element in the consultant's category is the preparation of construction documents. Construction documents must be prepared in order to define the work to be performed, to obtain all necessary permits, to secure various bid proposals, and to establish a clear contractual relationship between the company and the various vendors chosen to provide goods and services.

To protect the company's interests, construction documents must clearly and concisely describe all aspects of the work to be done. Contact the officer in charge of the facility to determine who will prepare the construction documents. (A budget amount should be allocated for the assistance of outside consultants if they are to prepare the documents.)

The landlord will often provide and/or pay for whatever construction documentation services are required for building standard construction. However, architectural, engineering and other services for tenant improvements which are not part of the landlord's original financial obligation are typically paid for by the tenant.

Professional consultants who provide architectural and engineering services for the facility should comply with the company's design and documentation standards. Sample workstation standards have been discussed in Chapter 3. Prepare them to suit the company's requirements, but remember that they should remain consistent with the design and space standards related to the company's particular criteria. Also, ask the prospective landlord if his architect or space planner and their consultants will follow the space plan and construction drawing formats which the company has developed after having reviewed the Construction and Furnishings Section.

If satisfactory construction documentation services are not available through the landlord, contact the company's own architect or space planner. It may be to the company's advantage to request a "planning allowance" from the landlord, while retaining an architect or planner independently, who will have the client's best interests at heart, with an allegiance to the company only.

Another major issue is procurement documents. Discuss with the company's architect, space planner or interior designer the best way to handle procurement documents for the job. Procurement documents for some millwork, furniture, furnishings, accessories, containers, signage and graphics may be prepared by members of the company's in-house staff. In such cases, have them follow the formats provided in this book and submit their documents to the Coordinator for review and approval. If the procurement documents are prepared by an outside consultant, include fees for such services in this category.

Miscellaneous and Unallocated Job Costs:

This is one of the most misunderstood, least used, yet most important categories in the budget. This is not a contingency. It is there to cover the multitude of items too small and too numerous to be line items on their own. In this category, enter about 2% of the total of all items recorded above. This 2% should cover that variety of minor costs attendant to any job which are too small to categorize individually, such as miscellaneous metals, special backing in partitions or walls, or relocation of a lighting fixture or air diffuser which is not working out in the originally designated location. This sum should also cover any small items which may have been overlooked in the major-category budgeting previously done, and any omissions from the budget.

Communications and Data Cable Systems:

The possible purchase and installation of new telephone equipment may be sizable. If the company is moving to a new facility, monthly service charges may be considerably higher than they are now. Be sure to discuss with consultants or in-house experts both the company's present and projected service needs and anticipated costs. This category should also include costs of material and labor for the installation of LAN lines, computer cable and any other forms of communications wiring.

Note: Contact the company's consultant or supplier for the appropriate estimate to use in this category.

Moving Costs:

Include the cost of relocating all existing equipment, office materials, files, supplies, and furniture that the company is planning to take to a new facility. Do not forget the cost of cartons, labels, tape, instruction and orientation forms, maps, and other packing or relocations materials. The cost of these miscellaneous items can add up to a significant expense.

Note: Ask at least two local moving companies to provide estimates for this budget. Preliminary estimates, although higher, can often be more accurate than actual proposals since the formal proposal may be artificially low (low-balling) to obtain more favorable consideration. Insert the higher estimate in this category.

New Equipment and Installation Charges:

This category is intended to include such items as new word processors, printers, typewriters, fax machines, copiers, dictation equipment, and the like, which may be required either due to added personnel or to equipment which is outmoded or not serviceable. Such equipment should be replaced at the time of the move. Also enter in this category all relocation and installation charges for copiers, offset equipment, and any other equipment requiring dealer installation when relocated.

Note: Contact the appropriate manufacturers for the figures that should be used as estimates for this category.

Administration and Travel:

Often the negotiation, design, construction and occupancy of a new or expanded facility requires travel by in-house or outside consultants. Staff members and the company's architects may wish to visit other office facilities or manufacturers of equipment or furniture. These visits often involve out-of-town and overnight travel. These costs should not be overlooked in the budgeting process.

Document Preparation:

The cost of producing copies of the various drawings, specifications, purchase documents, and procedures manuals is becoming significant these days; don't forget to take it into account.

Note: The company's architect or designer can help to develop this cost item.

Blank Spaces:

Enter here any budget items which have not been covered above, such as rent on space during construction, either as a result of negotiation with the landlord, or as overflow space required to accommodate temporarily displaced personnel.

Contingency (overrun):

Enter a number equal to 10%-15% of the total of all the above figures. As differentiated from the earlier *Unallocated Job Costs* category, this category is intended to cover such unpredictable additional expenses in a job as unusual, unanticipated inflation (anticipated inflation should be projected in each category) occurring during the period between the budget preparation date and the date when the bid is received; or such unforeseen circumstances as the default of a contractor, necessitating additional expenditures to complete the work, or extra rent costs caused by a delay in the completion of the construction process.

Total:

Enter all totals in this column.

Final Budget

After fully developing the Preliminary Budget Estimate, preparation can begin on the final budget. Follow the same procedure used for the preliminary budget. The principal differences between the two budgets are the level of detailed "back-up" material developed for each line item, and the timing of the Final Budget preparation. The Final Budget will be prepared from more empirical data than the preliminary. It will be based upon more finite programming, planning and design, and should represent a reasonable, accurate forecast of costs. This Final Budget, once approved, becomes the financial cornerstone of the project. To complete the project within budget, the Final Budget must be developed with great care.

For that reason, be sure to research each budget item with people from whom the company actually expects to be contracting for goods and services. A sample Final Budget form is reproduced in Figure 4.5. A full-size version of this form can be found in the Master Forms section.

To begin preparing the Final Budget, duplicate the Master Form. Then fill in each column after thoroughly reviewing the following detailed explanations.

The budget categories extending horizontally along the top of the budget should be filled in as follows:

Budget:

List the final budget for each of the goods or services listed in the left-hand vertical column.

$/SF:

Use this column to indicate the dollar costs per usable square foot of each item listed in the left-hand column, based on the total entered in the *Budget* column.

Vendor:

Use this column to indicate the actual vendors or contractors selected to provide each of the goods or services listed in the left-hand column.

Working with the Final Budget During the Project Development

Committed to Date (cumulative):

Use this column to indicate the exact total dollar amount which has been committed through signed contracts or purchase orders with the vendors listed for items falling within this category.

Final Budget

Facility: _____ Location: _____

Prepared By: _____ Phone: _____ Date: _____

Owner: _____ Contact: _____

Area	Estimated				Actual	
Rentable						
Usable						
Plannable						
Description	**A Budget**	**$/SF**	**Vendor**	**B Committed to Date**	**C Estimate at Completion**	**D Variance D=A+/-(B+C)**
General Construction						
Furniture and Fixtures						
Accessories						
Architectural						
Signage and Graphics						
Plants and Containers						
Artwork						
Packing and Moving						
New Equipment and Installation						
Legal						
Misc.and Unallocated Costs						
Tel., Data, Communications						
Travel						
Contingency						
Totals						
Authorized Changes						
Revised Totals						

Figure 4.5

Estimate to Complete:

Use this column to include a best estimate of additional dollar costs necessary to complete all items in this category. If it is estimated that an item will go over or under budget, reflect that estimate in this column.

Variance (plus or minus):

Use this column to indicate the variance in dollars between the total of the two previous columns immediately adjacent (*Committed to Date* plus *Estimate to Complete*) and the *Budget* column.

This Variance column will enable all other interested parties to monitor the financial status of the project on a continuous basis.

$$A = (B + C) + / - D$$

or

$$D = A + / - (B + C)$$

The budget categories listed vertically along the left-hand side of the final budget form under *Description* are generally the same as the corresponding budget categories along the left-hand side of the Preliminary Budget Estimate form. Note that there are more blank lines under the *Furniture and Fixtures* heading. These lines are included because it is possible that more than one outside vendor will be used. Use a separate line for each vendor from whom furniture and fixtures are being purchased.

Likewise, there are blank lines under *Architectural and Other Outside Consultants*. Again, these lines have been inserted because more than one outside consultant may be used. Use a separate line for each outside consultant.

There are two line items at the bottom of this column. The first is *Authorized Changes (Cumulative)*. After the final budget has been approved, it should not be exceeded, except where changes in the program initiated by the officer in charge of the facility occur. Be sure that whenever anyone initiates a change in requirements, they are advised in writing of the cost implications of the proposed change. If the change is approved, enter the cost in this column.

The *Revised Total* column should only be used if authorized changes have been entered above.

Be sure to carefully research with vendors each item listed in the Final Budget. For example, when preparing the budget for furniture and furnishings, do so by making a precise takeoff of components the company will require for its new facility, using the information found in the *Furniture* Section. (A takeoff is accomplished by measuring exact quantities as shown on plans.) Work with a local contract furnishings dealer to help develop the budget.

The Final Budget form is designed for continued use throughout the project. Through regular updating, it will be possible to keep yourself, as project or facility manager, and management fully apprised of all changes in budget status. Review the budget no less often than monthly. We would strongly recommend updating this budget every week during the organizational and bidding process. Do not hesitate to contact the company's architect and other consultants for information on the budget control process. Architects, in particular, are likely to have had years of experience in developing and maintaining budgets.

Chapter 5

SCHEMATIC DESIGN

Introduction

Before beginning the schematic design process, the company should complete all the work described in the previous chapters. It is important that accurate floor plans of the space to be occupied be available, whether it is the current facility (with or without additional space) or a completely new facility.

Once accurate floor plans are in hand, the schematic design phase may be started. Begin by assigning each department to its most appropriate location in the facility. Evaluate adjacency requirements between and within departments, and then determine actual furniture placement for each workstation and common area. This chapter explains how to accomplish these tasks.

Two basic types of office plans are in use today: a *conventional* office, which separates work functions by means of full-height walls and doors, and an *open-plan* office concept, with few or no full-height walls. In the open-plan concept, workers are usually divided from one another by workstation systems furniture panels or screens. It is likely that a new office plan will utilize both open-plan and enclosed offices.

One of the major disadvantages of a conventional office is that it is both difficult and expensive to re-configure. In an open-plan or semi-open-plan office, by contrast, systems furniture panels and screens can be moved easily and inexpensively. One of the major disadvantages of an open-plan office, however, is that it is often difficult to discuss highly confidential matters in private. For these reasons, most companies will find the best path to be a design approach that is essentially open-plan for flexibility, but uses full-height walls and doors for privacy or status. Providing a number of small conference rooms that are easily accessible for confidential meetings and/or high concentration work is a practical solution to temporary privacy or distraction problems.

Finalizing Space Standards

At this point two prerequisites should be complete: the study of workstations and other space element standards, and the Space Requirements Questionnaires.

Standardization of specific workstation configurations will have begun to create some basic design concepts. By having those standards produced as a result of in-house evaluations, you will have progressed a long way towards their universal acceptance by all personnel. For facilities managers, this will help eliminate the time and expense of developing new designs each time you plan, or re-plan a facility.

As workstation standards are developed, it is important to have a number of discussions with the company's senior management. Management usually is focused on producing the company's products. Management rarely has the time or opportunity to consider that how they manage and how day-to-day operations are conducted becomes a part of the company's corporate culture. This corporate culture will very likely have a direct bearing on the type of facility that will best accommodate management's business style. How this facility functions and looks, what kinds of personnel it employs, and what their unique needs are, are subjects not often considered when trying to get the product to market. It should not take senior management long to recognize the interrelationships between their office facilities and work product. Assist them by offering alternative concepts which will produce the most efficient and responsive work environments and the most practical use of occupancy costs. How open-plan workstations are combined with so-called conventional or closed office design will be determined in the company's basic planning and design concepts.

Select various, but compatible, lines of systems furniture with appropriate components. All seating, free-standing furniture, carpeting, fabrics and color schemes should meet the basic design concepts and standards which suit the company's operations and image in this particular building.

Each individual's workstation will have its "package" of elements, e.g., systems furniture panels, work surfaces, components, files, seating, task lighting, accessories, etc. There may be a number of alternate suppliers for many of these elements or selections, although the company may have limited selections to one or two. In our sample standards we have provided a sample list of the elements comprising the typical workstation, with the exception of the systems furniture panels or walls. Also, we have assigned a furniture code to each furniture type. This is a convenient way of designating a *package* for each workstation.

Policy

Once final standards are determined, give serious consideration to approving and distributing a written policy concerning these standards and their application. By this point in the project, the Workstation Assignment matrix should be complete, and specific workstations should be assigned to all company personnel. How the company elected to apply or assign the standards is again, a matter of corporate culture and operating style. The assignments may have been made on the basis of job functions, job titles, salary levels, arbitrary assignment by management at the time of planning, or combinations of the above.

The next question is how to relay those policy decisions to the company's personnel; or whether to relay the information to all personnel at all. The company may wish to merely provide a policy statement such as the one which follows, or they may wish to include the workstation matrix which was prepared when workstations were assigned to each individual. There are a number of ways of writing policy statements reflecting the results of the basic concepts. Figure 5.1 is an example of the material such a statement might contain.

Blocking Plans

Careful placement of departments within the space will improve work flow and efficiency. Therefore, the first step in the schematic design process is to prepare adjacency layouts – that is, to decide which department goes where in the new facility. This can be done as follows:

1. Review each department's response to the *Space Requirements Survey* question about adjacencies.
2. Interview each department head, section head, and branch manager to discuss which departments and sections "must be" (essential) next to one another and why, and which "would like" (important or convenient) to be next to one another and why. Ask each department head to make a list of optimum adjacencies for his or her department/section, numbering each department and section on the list in descending order of importance. Try to determine if the need for adjacency can be dealt with by phone, fax, etc., or if the need is merely a matter of habit.

   It is important to recognize that some adjacencies need not be directly related to reporting structures. It may be perfectly valid for a supporting department to be in a totally different building or area from a department to which it supplies the major portion of its service. Many bank check processing departments are located miles away from their main offices.
3. In order to decide which departments should be placed next to one another, prepare an adjacency matrix based on the information gained in Step 2 above. An adjacency matrix simplifies the process of evaluating adjacency requests by putting them all on one sheet of paper. Figure 5.2 shows an Adjacency Matrix form in reduced size; a full-size form can be found in the Master Forms Section.
4. The Area Requirements By Department form and the Facility Area Requirements form completed during the Space Requirements Survey should be available for reference.

Standards Policy Statement

To all personnel of The Company, Inc.:

As you know, we have acquired new office space, into which we will relocate. We have also been reviewing all of our job functions and what each of our work spaces should be like and how they should be furnished and equipped. We have made every effort to accommodate each individual's needs and desires. Based on these analyses, we have designed a set of standards which we believe will be both functional and flexible, and which we intend to use in our new facility.

The purpose of these standards is to provide consistency of facilities and appropriateness of work areas to job functions for employees throughout The Company and all affiliates and subsidiaries.

Due to variations in the assignment of titles throughout the company, salary grade levels have been used as the primary determinant of work area assignment. This method of allocating office types will not be without exception; therefore, "Functional Override" provisions may accommodate particular requirements. A department head may request such an override which must be approved by the Officer in Charge. However, as with any standards system, the key to success is consistent application with the fewest possible exceptions.

Figure 5.1

To use this matrix, list all departments twice, once vertically and once horizontally, each time in the same order. (Figure 5.3 shows a completed example.) Start each list in the upper left-hand corner with the same department. Using the code on the form, note all adjacency requests. Ignore the shaded squares running diagonally down the page. They represent those points on the graph where each department intersects with itself.

Although each department is listed twice, record its adjacency requests only in its horizontal listing. Do so by filling in the appropriate boxes on the horizontal line to the right of the department's name. When the matrix is finished, the two triangular halves of the matrix on either side of the shaded squares are not usually mirror images of one another. This is because department heads may not agree with one another on adjacency needs. Nevertheless, if this does occur, it might be wise to double check your information. There could be an error.

Note that elevator lobby, central files, lunch room, etc. are listed. This is because certain departments need special access to areas which are not specifically departments. It is important to have this information early in the planning process.

Look for groups or clusters of requests which have been characterized as essential (E) or important (I). It will be necessary to re-do the form two or three times before clusters become apparent. Each time you re-do the matrix, adjust the list of departments so that departments requesting essential (E) adjacencies are listed next to one another. As this process is repeated, the clusters will fall into more recognizable groups. Tight clusters become excellent departmental groupings for floor-to-floor distribution of departments.

As can be seen in the example in Figure 5.3, the same department (Administration) starts both lists in the upper left-hand corner. The matrix has been redone a few times prior to this stage, and several departmental clusters have become apparent.

Also note that in several instances department heads did not agree. For example, Audit says it is essential (E) that it be next to Data Entry. But Data Entry makes no mention of Audit at all. Verify the information with both departments. If the information is correct, try to solve Audit's perceived need.

Plannable Square Foot Areas

Before any further work can be done, determine the plannable square footage on each floor to be designed. Remember that the usable/plannable square footage is somewhat smaller than the amount of square footage on which the company is paying rent. Often the building owner or the owner's architects will provide usable square footage figures for the floor(s) you are leasing. However, it is imperative that you check any figures you may receive.

Under the BOMA standard of usable square footage, certain areas of the floors may be included which will not be "usable" for space planning purposes. Such areas are toilet rooms, telephone and electrical closets, convectors along outside walls, columns, cross-bracing, etc. These areas are all considered a part of "usable" under the leasing definitions. It's tough to put a workstation in a telephone closet!

Adjacency Matrix

Facility: _____ Location: _____

Prepared By: _____ Phone: _____ Date: _____

E = Essential that these two departments be adjacent
I = Important that these two departments be adjacent
C = Convenient that these two departments be adjacent

Area	Estimated	Actual
Rentable		
Usable		

Figure 5.2

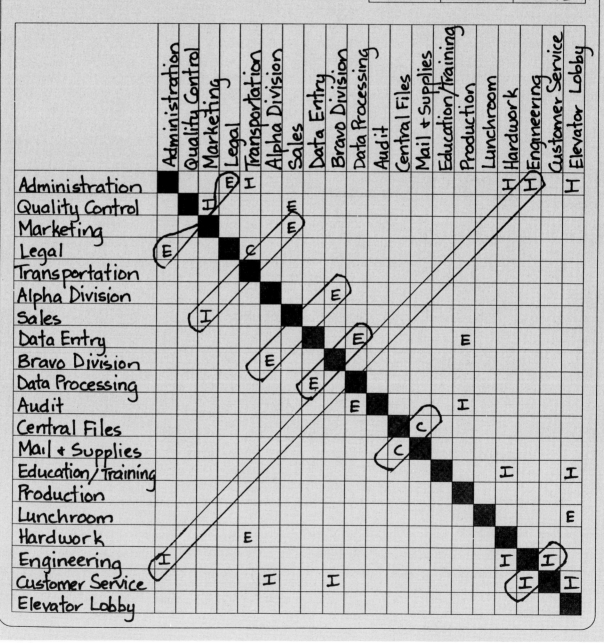

Figure 5.3

To calculate the plannable areas on full floors, obtain a copy of the floor plan from the building's management or the building architect. Shade out the building's core areas (elevator, stair shafts, restrooms, mechanical areas, and the like), since they are not space within which you can plan office space usage. Figure 5.4 is a sample of such a plan.

Divide the floor area into a grid as shown in Figure 5.5. Do this by connecting the center points of the columns within the floor's gross area.

EXAMPLE

In the sample floor, we broke the space into 24 bays, each of which is 25' on a side.

25' x 25' = 625 SF

Each bay therefore contains 625 square feet.

24 bays x 625 SF = 15,000 SF

The gross area of the floor is then 15,000 square feet. Most buildings will not be as modular as this sample floor, so be sure to measure carefully, and don't forget to add (or subtract) the area that lies between the outside column lines and the face of the window.

To determine the plannable area, add up the square footage taken up by the core areas, columns, air-conditioning or heating units at the window wall, window sills, and any other elements that render floor space unusable; then deduct the total unusable square footage from the gross floor area. (Be sure to base your calculation on the scale at which the floor plan has been drawn. Most drawings will be at 1/8" scale, which means that 1/8" on the drawing equals 1' on the actual floor. To make your measurements easier, use an architect's scale, which can be purchased at most office supply stores.)

The plannable square footage for this sample floor has been calculated as approximately 13,635 SF.

Study the plan of each floor the company will occupy. If there are any differences in perimeter shape, dimensions, or core area configurations, use those measurements and the procedure described above to determine that floor's plannable area.

When leasing a partial floor, calculate the plannable square footage only within the area to be leased.

Stacking Plans

If the facility has more than one floor, the next step is to determine which departments should go on which floors. This process is known as *stacking* or *vertical adjacencies*. For smaller projects, determining stacking is a relatively simple task. In larger projects, however, there are a number of factors which must be kept in mind. These include: the desirability, or lack of desirability, of public access to each department; requirements for access to special services such as security, mail room, supply rooms, receiving areas and freight elevators; and the need to facilitate employee access to such common areas as lunchrooms and dining rooms, dispensaries, and education and training facilities.

After the adjacency matrices are completed, it is time to develop stacking plans. The stacking process requires experimentation. Begin by listing the plannable square footage available on each floor. Next, assign the department groupings which evolved during the adjacency studies to specific floors. Keep two factors in mind when assigning each department: its square footage requirements, and its adjacency requirements.

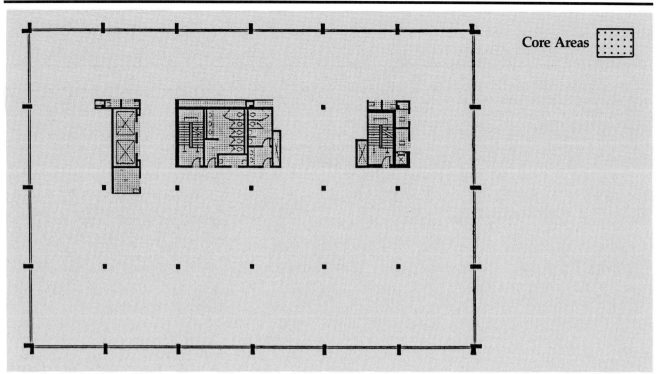

Figure 5.4

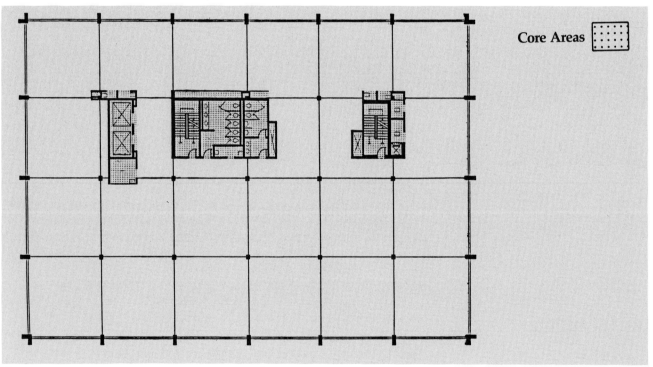

Figure 5.5

It is important to assign departments to a floor in such a way that their total projected future square footage requirements will not exceed the total calculated usable area for that floor. In practice, however, departmental totals sometimes exceed floor totals. If the excess on a floor is not greater than 300 or 400 square feet on a 15,000 to 20,000 square foot floor, it is possible to accommodate those departments through design efficiencies by making minor adjustments at the schematic design stage. However, if the total excess is greater than 500 square feet, consider shifting a small department or part of a large department to another floor.

Attempt to assign department locations in such a way that their adjacency requirements are met to the greatest extent possible, making physical communications lines drawn between departments as short as possible. The question of physical adjacency is a very sensitive one. There is much to be said for such technologies as E-Mail and LAN computer systems, reducing some need for face-to-face communication. On the other hand, studies by Duncan Sutherland indicate that the chances for weekly communication drop from a 25% likelihood at about 15 feet to less than a 5% probability at about 20 feet, and that after 100 feet the probability of weekly or even monthly communication is nearly lost.

When locating and stacking departments whose adjacencies are not critical, group smaller, easily relocated departments adjacent to departments which are likely to experience volatile growth. It is important to try to provide a growth area in the form of smaller, easily-relocated departments next to each department which is forecast for rapid growth. This practice is second in importance only to placing departments next to one another when they have expressed an essential need for an adjacency to one another.

Once stacking the departments is completed, obtain management's review and approval. The work should not continue until stacking diagrams have been approved by all required parties.

Figures 5.6 through 5.9 are an analysis of a five-floor facility (basement and four floors). Figure 5.6 shows the gross and plannable square footage figures for each floor. Figures 5.7, 5.8, and 5.9 are worksheets showing a proposed stacking plan for the five floors. The person planning the facility found it necessary to adjust the square footage assigned to the four departments planned for the basement. Also, an allowance of almost 1,000 square feet was made on the first floor to cover ground floor entry and areas.

Once the stacking plan has been approved, it is time to begin creating final block diagrams, which allocate specific floor areas to specific departments. Using the adjacency matrix clusters previously developed, create a number of alternative block diagrams locating all departments in the floor space. If there are shared common spaces, that is, spaces used in common by two or more departments or sections, such as reception areas, file areas, conference rooms, libraries, copy or vending stations, etc., treat those areas as a unique section or department.

Consider Departments A, B, C, D, and E. They have all been assigned to the same floor in the sample facility. An adjacency matrix for the five departments is shown in Figure 5.10.

An adjacency block diagram allocating specific floor space to the five sample departments is shown in Figure 5.11. Note that Department E found it essential it be near an elevator, but did not find it essential to be near any other department. Therefore E is an appropriate department to fill the open space on this floor.

Preliminary Stacking Plan Work Sheet

Facility: **Branch**   Location: **Fairview, IL**

Prepared By: **David Johns**   Phone: **555-8989**   Date: **11/22/93**

Floor		Department	Square Feet	
Number	Gross S.F.		Per Dept.	Plannable
Basement	15,000			11,988
1st	15,000			13,397
2nd	15,000			13,635
3rd	15,000			13,635
4th	15,000			13,635
	75,000			66,290

Notes :

Figure 5.6

Preliminary Stacking Plan Work Sheet

Facility: __Branch__   Location: __Fairview, IL__

Prepared By: __David Johns__   Phone: __555-8989__   Date: __11/22/93__

Floor		Department	Square Feet	
Number	Gross S.F.		Per Dept.	Plannable
Basement	11,988	Central Files	3950	
		Mail + Supplies	1935	
		Education + Training	2030	
		Lunchroom + Lounge	2250	
				10,165
1st	13,397	Customer Service	1375	
		Alpha Division	7470	
		Bravo Division	3581	
				12,426
				22,591

Notes : _____

_____

_____

_____

_____

Figure 5.7

Preliminary Stacking Plan Work Sheet

Facility: __Branch__   Location: __Fairview, IL__

Prepared By: __David Johns__   Phone: __555-8989__ Date: __11/23/93__

Floor		Department	Square Feet	
Number	Gross S.F.		Per Dept.	Plannable
2nd	13,635	Audit	1012	
		Data Processing	5655	
		Reproduction	938	
		Production	4138	
		Data Entry	977	
				12,720
3rd	13,635	Quality Control	2713	
		Marketing	5725	
		Sales	5288	
				13,726
				26,446

Notes:

Floors 3 & 4 could be reversed

Figure 5.8

Facility: __Branch__   Location: __Fairview, IL__

Prepared By: __David Johns__   Phone: __555-8989__ Date: __11/23/93__

Floor		Department	Square Feet	
Number	Gross S.F.		Per Dept.	Plannable
4th	13,635	Administration	1594	
		Legal	2494	
		Admin. for office	300	
		services		
		Transcription	1219	
		Hardwork	2012	
		Strategic Planning	1831	
		Engineering	3818	
				13,268

Notes :

Consider shifting :

a.) Admin. for office services to 2nd floor

b.) Audit from 2nd to 4th floor

Figure 5.9

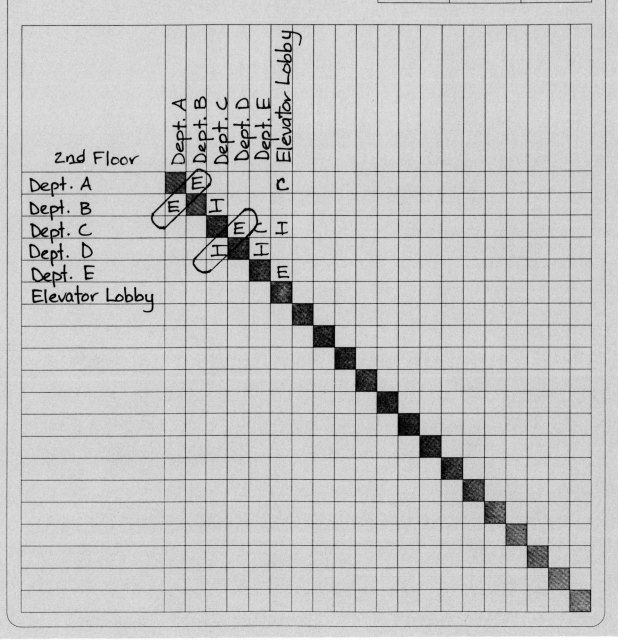

Figure 5.10

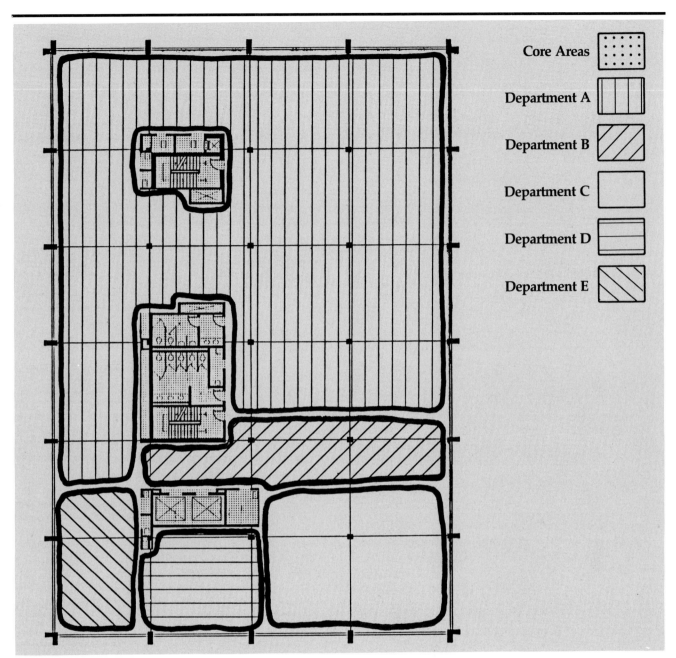

Figure 5.11

While there are a number of shapes that the blocks could take, every attempt should be made to avoid long thin shapes that spread out the department, and odd shapes that could make intra-departmental access and communication difficult. The purpose of the planning process is to improve work flow and communication within departments, as well as between them. To aid in doing this, try to visualize how the furniture could actually be placed within each block as it is drawn.

When all the block diagrams which represent reasonable alternatives are completed, present them to management. Based on subsequent discussions with management, continue experimenting and refining until there is agreement that an optimal adjacency plan has been developed.

Keep in mind that adjacencies are vertical as well as horizontal. That is, a department will find it relatively easy to communicate with those departments on the floor above it and on the floor below it. If heavy traffic between floors is anticipated, for the duration of the lease, consider using the existing internal fire stair if local building and fire codes will permit. If not, adding an inter-floor stair could be worth the added expense.

The impact of block diagrams on a stacking plan can perhaps be best appreciated by looking at the "exploded" drawing in Figure 5.12. The drawing shows block diagrams for a sample five-floor facility, and how each floor relates to the ones above and below it.

Basic Design Concepts

In order to create and maintain a consistently high level of quality and continuity of planning and design philosophy as a function of the corporate culture throughout all the company's facilities, the same basic design concepts for all projects should be developed and followed. If office space needs to be utilized in diverse ways within the company, it may be necessary to develop more than one concept. A particular facility may need mostly enclosed private offices, while another facility might be nearly all open plan. What is important is that the basic concepts have given consideration to what it is the company does and how its employees do it.

An examination of the company's style of doing business and the various job functions within it will often disclose great similarities in how work is performed regardless of what the job title may be. The corporate comptroller and the director of marketing may have totally different titles, job descriptions and duties, but may need very nearly the same workstation. A coding clerk and a word processing clerk may also be performing completely different jobs, but require nearly identical workstations.

It is not only important to define the company's basic planning concepts, but to also review the design style best suited to the corporate culture. Some companies see themselves as basically "high-tech," utilizing very contemporary furniture styles and intense color schemes. Other companies see themselves as more conservative, using softer color schemes and more traditional furniture and furnishings.

Even within a single facility, the design concept may range over a spectrum of styles and color palates, each dependent on criteria such as departmental or area function.

Once definition efforts have been concluded, a fairly well-defined basic planning and design concept should have been generated. The concept should try to be broad enough to cover the company's various needs and functions, but sufficiently definitive and limited to qualify as a policy statement. A good planning and design concept should include:

Example:

Sample Floor Stacking Diagram

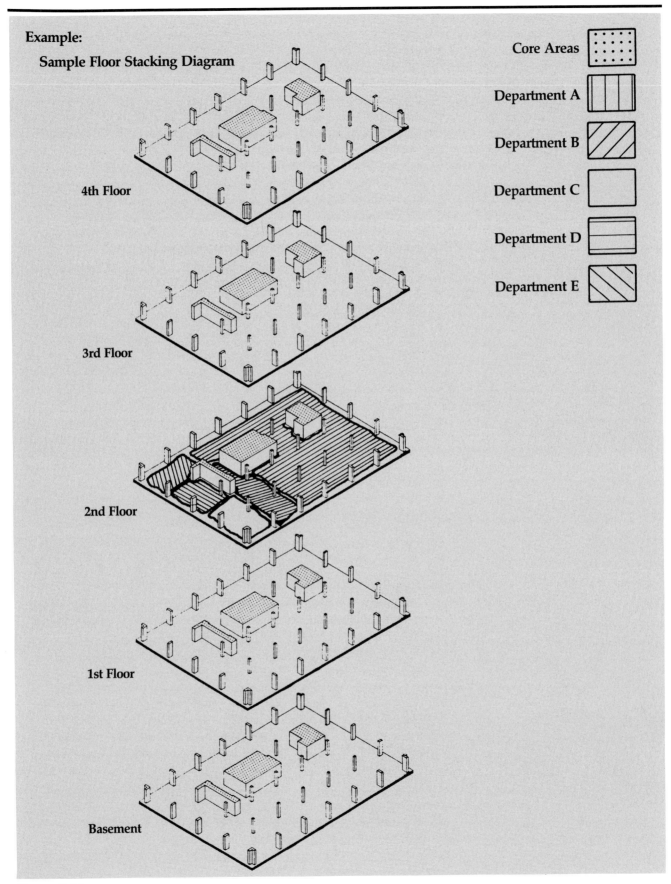

Core Areas

Department A

Department B

Department C

Department D

Department E

4th Floor

3rd Floor

2nd Floor

1st Floor

Basement

Figure 5.12

- A basic planning approach
- A basic design concept
- Specific workstation modules and types
- Furniture and furnishing standards
- Color and fabric selections

A basic planning and design concept makes the task of planning and designing a new or remodeled facility significantly easier. In addition, the basic design concept, when properly followed, will ensure a design direction which will lend continuity to all the company's facilities.

Design Development

At this point most of the so-called *standards* questions have been addressed; the various departments are located in specific places on specific floors; and a design concept has been chosen. It is time to begin the actual design. This phase of the design process is at the schematic level because various components of the facility must still be isolated and organized.

The first step in this phase is to plan for inter-departmental adjacencies, both *primary*, or the most critical, and *secondary*, or those less critical.

Inter-departmental adjacencies refer to the relationships of work groups within a department or section. Large departments (such as Department A in Figure 5.10) have numerous work groups. These must be located within the departmental block with even more care and consideration for work flow and communication than was given to the placement of departments during the stacking and blocking steps previously described.

One possible solution is to place those departmental work groups and functions which require full-height walls or enclosures near the building core. If this is not possible, clustering groups with high-wall requirements together in appropriate functional locations is another solution.

Circulation

The idea of *major circulation* or *primary circulation* refers to the need to allow room for people to move about from one area to another. Circulation concepts and basic circulation requirements differ from one building to another and depend, among other things, on the distance between the building core and exterior walls. When the distance from core wall to exterior wall is 25 to 35 feet, often only one passageway or corridor that is directly adjacent to the core area is needed. If, however, the distance between core wall and exterior wall is 45 to 60 feet, two passageways or corridors running parallel to one another, usually on opposite sides of the core area, may be required. (Figure 5.13 shows both conditions.)

Major circulation patterns should connect building entries, elevator lobbies, stairways, core functions, restrooms, major departments, and common use areas (reception rooms, conference rooms, copy and file rooms, and the like) in a simple, clear, and logical fashion. Because circulation paths offer the means for leaving a building in an emergency, there are a number of building codes which regulate circulation planning. Be sure to consult local authorities during the planning process. Also, be sure to read the section *Conforming to Codes*, which follows this section.

Secondary circulation refers to the areas provided for access to and around individual workstations, work clusters, and special equipment. Here the five departments have been divided into major work groups. Primary circulation patterns have also been established.

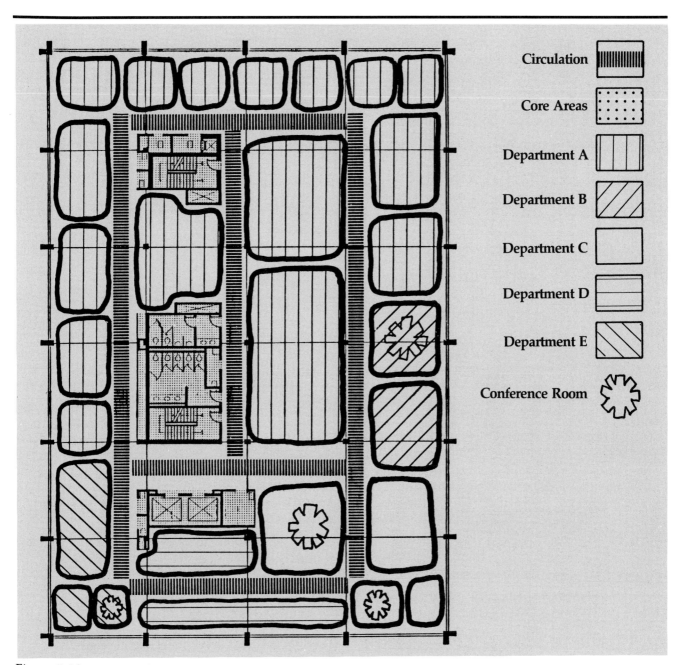

Figure 5.13

After management's approval of the adjacency and circulation plans is obtained, break down the departmental work groups into actual workstation areas. Cluster these by size and type, supervision requirements, and relationship to special functions and common use areas. Thoroughness in this step of the planning process will greatly simplify the actual detailed space plans which follow. This step can also be a valuable time-saver, as it allows the planners to get down to specifics without having to confuse the drawing with a great deal of detail. Consequently, this step greatly simplifies the departmental and management approvals process.

Carefully consider the information gathered from department heads about specific work storage, work distribution, mail and supply delivery, mail pickup, and any other key operational factors requiring equipment that takes up floor space. Block out areas for these items, as they must be included during this stage of the planning process to ensure that they aren't forgotten later.

Keep in mind that some workstations or mini-conference rooms will require more privacy than others, according to the work or task being performed. Such areas may require special locations or acoustical panels or screens with heights greater than those of standard workstations. These high-screen areas can be strategically placed to provide visual breaks in large open areas wherever possible and appropriate.

When all workstation types, common use areas, and pieces of equipment have been located, the most desirable plan of all the alternatives evaluated should be presented to department heads and management for review and approvals.

A lot of work has been done to get to this point in the planning. The detailed design phase which follows this phase requires a great deal of precision and is not easily modified. It is essential that there be a very careful review of the inter-departmental adjacencies with the officer in charge of operations in this office – The Boss. This person should carefully examine how departmental relationships, sizes, staffing and anticipated growth have been interpreted. After some review and thought, a number of major changes may be wanted.

In the drawings in Figures 5.14 and 5.15, each department has been subdivided into actual workstations. Notice how the workstations of some work groups have been drawn at an angle to increase the length of the workstation chain along the center line of the group (spline), and to add visual variety to the plan.

Conforming to Codes

In most local jurisdictions in the United States, construction or alterations exceeding $100 in value require a permit issued through the local building authority. Jurisdictions may be either city or county, and the issuing authority may be called the Building Department, Bureau of Building and Safety, Bureau of Building Inspection, or other similar title. It is very likely that the company's plans may have to be "stamped" with a registered architect's seal. This requirement anticipates that the architect has inspected the plans and that, as a registered professional, he has approved them. It is also possible that some of the other plans, such as electrical or mechanical, may have to bear their respective engineer's seals.

Prior to initiating the planning process, a visit should be made to the local building authority for a complete review of the approving agencies having jurisdiction over the project. In most cases, the agencies will include the building authority itself, the fire department, and the health department when any food handling or food services are involved.

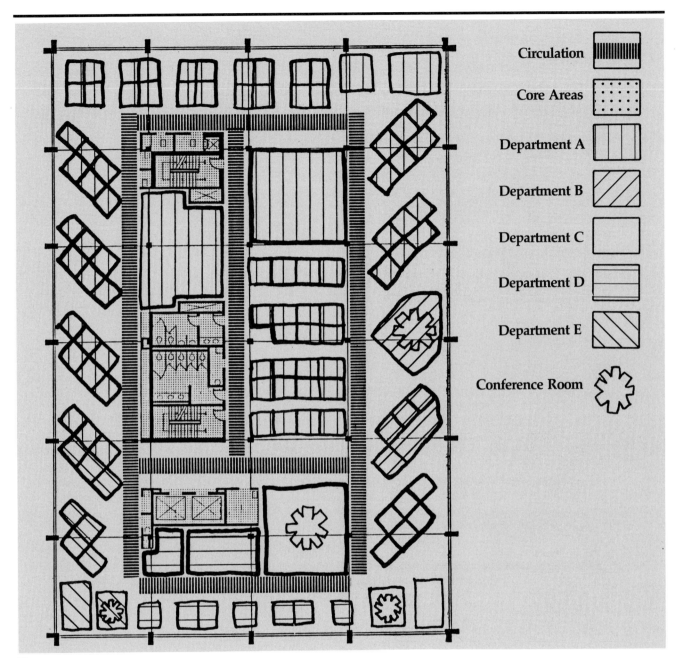

Circulation

Core Areas

Department A

Department B

Department C

Department D

Department E

Conference Room

Figure 5.14

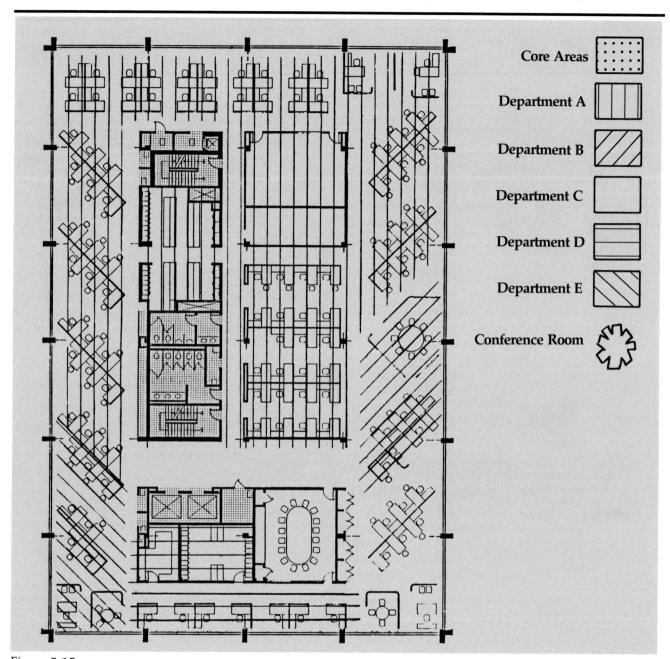

Core Areas

Department A

Department B

Department C

Department D

Department E

Conference Room

Figure 5.15

In most cases, the review by each of these agencies is coordinated by the building department.

At the preliminary meeting with the building authority, try to get answers to the following questions:

- Which agencies will have jurisdiction over the project?
- How must construction documents be submitted (and in what quantity) to these agencies for their review and approval?
- Will the agencies provide a preliminary plan check or review service prior to a formal application for a building permit, and will they provide and honor preliminary approval on plans?
- What fees are required for plan checks and permit processing?
- How much time is required for the plan check and permit issuance cycle?
- What codes, ordinances, and regulations (federal, state, county and city) apply to the project?

The information gathered at these early meetings will have a major effect on scheduling, budgeting, and the planning process.

Particular attention should be paid to the chapters in the locally applicable building code which pertain to exiting and to the occupancy types to be contained within the facility. Most of the planning the company will use will probably be considered *open planning*. However, most codes are written for conventional offices with walls, corridors and doors. Consequently, it is important to learn how local plan check officials interpret open plans from an exiting standpoint.

The portion of the building code dealing with exiting should be read carefully and summarized before starting the planning process. Building codes differ throughout the country and are often modified by local regulation. They are always riddled with cross-references, exceptions, and other confusing material. The building code sections pertaining to occupancy types must also be carefully read and summarized.

The occupancy types most commonly encountered in office facilities are "office" and "assembly." Large areas used for dining, lunchrooms, training rooms, or large conference rooms might be considered "public assembly" areas which may require fire separation (a fire-resistant wall), special emergency lighting, etc. An initial survey of the requirements for the facility should be discussed with building officials to determine the occupancy types.

Occasionally, the maximum size of office areas may be restricted because of the construction type of the building to be occupied. Most new office buildings are Type 1 (totally non-combustible construction) and will have very few restrictions relating to area separations. Once a building has been selected, discuss the question of basic restrictions with the local building official.

Building officials typically will be most concerned with exiting patterns within open plan areas, distances to exits, widths and configurations of corridors, fire separations between areas and corridors, fire ratings of doors, and provisions for the disabled.

The fire department will be concerned with exiting patterns, exit sign locations, emergency lighting, fire separations and fire suppression systems.

The health department will usually be concerned with finishes and materials used in food service areas, cafeterias, dining areas with food preparation, and provisions for food service employees.

It is most helpful if preliminary plan reviews take place with the building authority at the following times:

- Upon completion of block diagrams, prior to planning and scheduling; or
- Upon completion of space planning, prior to the final review with each department, and budgeting, and scheduling; and
- Upon completion of final space plans, prior to the commencement of the construction documents. At this stage an attempt should be made to obtain a preliminary approval of the plan for exiting purposes.

Note: It is not very likely that you will be able to achieve any of the previous reviews due to the heavy workloads building officials have to carry. They may brief you on which edition of the Building Code they are following and what special local requirements they may have.

During the plan review process, it is important to remember that the plan check personnel are not always the same people who initially reviewed drawings, and they will not necessarily honor approvals made by other building officials.

It is also important to remember during the construction period that the field inspector is rarely the plan checker. The field inspector may not honor approvals made earlier by plan checkers and has the obligation to uncover areas of noncompliance to codes which the plan checker may have missed. Surprises cannot always be avoided, but attention to codes and their local applications is important during all phases of the planning, design, and construction documentation for the project.

In some cities the process of plan checking is quite complicated, with special personnel assigned to speciality areas such as structural, mechanical, electrical, egress, food service, life-safety, etc. In these cities it is quite common to employ special consultants to assist the architect or owner in submitting plans for review and approval. It is wise to inquire as to the practice in the city where the new facility is planned. If it is common to employ such a consultant, consider employing one for your plan submission. They can save the company many days, if not weeks, in securing approvals. Delays in plan approval can destroy the best schedules.

Designing for the Disabled and ADA

Under Federal law (the Americans With Disabilities Act), all areas available to the public must be accessible to and usable by individuals with disabilities, with facilities appropriately modified for their use. While this may seem to represent a substantial allocation of funds, design effort and space, consider that at any given time ten percent of the U.S. population has some form of disability, whether temporarily so (a broken leg from a skiing accident or the like), by virtue of age, or by some form of permanent disease, or physical or mental impairment. If a facility is not designed to be fully sensitive to the problems of accessibility to this segment of the population, a large pool of potential employees and customers is being ignored, and the company is not complying with the law.

When designing the company's office space, keep in mind the requirements of individuals with sensory and cognitive impairments, as well as those with mobility disabilities. A good starting place should be in the parking lot or structure, or at curbside. Trace the necessary route for ingress and egress. Are there adequate handicapped parking spaces? Are curb ramps in convenient locations, at proper slopes, and not so smooth as to be a hazard in themselves? Are entry doors equipped with the proper pulls, and not too heavy to open? Are doorways wide enough when measured from the door when it is open at 90 degrees? Should entry doors be power-assisted? Are entry vestibules deep enough to allow for wheelchair maneuvering? Are ramps designed with adequate space for turning a wheelchair; and are there level spaces at sufficient intervals to provide rest stops? Restrooms should be specially equipped and laid out, and elevator call buttons, water fountains and pay telephones placed at specified heights and designated with Braille characters.

For the visually-impaired, there should be a change in the floor texture in front of any "danger" areas (such as the tops of stairs, at curbs, or in front of escalators); and important signage information should be given in Braille. There should not be any overhangs low enough to permit someone to hit their head on them. Designing for the visually and hearing-impaired includes installing both audible and visual alarm signals. Aisles need to be adequately wide and clear of obstructions. Thresholds should be tapered and not too high, and carpeting should be of a proper density and pile height so as not to create an unnecessary obstruction. These are just a few of the components required in designing to accommodate individuals with disabilities. We have not tried to give specific recommendations as to heights, widths, angles, etc. Refer to the Federal regulations for more information. The interior designer or architect should be able to provide data on Federal and local requirements.

Any existing structure the company leases must also comply with the Federal regulations. If it does not, part of the lease negotiations should include the modifications necessary to meet current standards. If a structure is to be remodeled for tenancy, regulations require that the facility be altered to accommodate those people with disabilities. For a full discussion of requirements and recommendations of design for the handicapped, refer to:

The Americans with Disabilities Act (ADA)
Accessibility Guidelines for Buildings and Facilities
Effective January 26, 1992
1-800/USA-ABLE

Uniform Federal Accessibility
Standards FED-STD-795
April 1, 1988
GSA, Department of Defense,
Department of Housing and Urban Development,
and U.S. Postal Service

Write to :

ATBCB, 1111 18th St. N.W., Suite 501*
Washington, D.C. 20036-3894
(202/653-7834 V/TDD)
(202/653-7863 Fax)

*Architectural and Transportation Barriers Compliance Board,
 "ACCESS Board"

Interior Design   It is at about this point in the project that the company will need to turn over more aspects of the planning to professional designers. If the design work will be performed by an outside source, spend some time investigating the design community available. Try to avoid being led into an "old boy," or "old girl" connection. Such a connection that you may be forced or pressured into using may very well be highly qualified, creative, easy to work with, and competitive in their fees, but do not necessarily count on it. Try to make your own investigation and arrive at your own conclusions and decisions or recommendations.

There are basically two principal design professionals: *architects* and so-called *interior design firms*. Within each there is likely to be a large amount of qualifications overlap. Most architectural firms employ a large number of designers who, although not registered architects, have varying degrees of capability and experience, and who do most of the work. Many, but probably not most, interior design firms have registered architects on their staff. Either may be capable of performing high quality work. Both are capable of doing poor quality work. Whichever way the company decides, look for the same qualifications and expect the same high quality results.

Some furniture dealers may hold themselves out as being qualified designers. Few, if any, have registered architects on staff, nor are they likely to be qualified in building codes. They may suggest that they will do a design "for free." Whatever design services they provide have to be paid for in some manner. Usually those costs become a part of the dealer's markup. In addition, furniture and furnishings selections will be limited to those manufacturers which the dealer carries. However, most dealers representing major systems furniture lines have the capabilities of performing detailed furniture takeoffs to determine necessary quantities.

Qualifications

There are at least five basic qualifications to look for in a designer. It is possible to forego one or perhaps two of these basics in order to secure a very special talent. However, do not let the glitter distort the final selection. Remember two things: this will be the company's facility, not the designer's; and when the job is finished and the building occupied, it will be without the designer. He or she will not be sharing the results of your joint endeavor.

Programming and Planning

Can the prospective designer demonstrate a clear understanding of the value and process of generating the space requirements, space standards, interior programming and adjacency procedures? Or, does he or she anticipate coming to the department interviews with no structured procedure or outline?

Experience

Is the prospective designer anxious to show off recently completed projects of similar size and scope, to introduce their client contacts, and to discuss that client's objectives and the design solutions? Is the client happy not only with the result, but does the client feel it received reasonable value for the design fees and the cost of the facility development?

Creativity

The ability of a designer to be creative is measured not only by how well a space may function and appear when completed, but how well the designer sought out the client's concept of its culture and image, and how well the design captured the desired feeling. Creativity is also measured by how well the designer is able to preserve portions of the existing facility and furniture which you agree are worth preserving, and to reuse those elements.

A word of caution, here. When the company employs a designer, it is employing not a mere technician and pencil pusher, but an educated and creative person. The temptation of many project management people, coordinators, etc. is to over-direct the design team. The management of good, creative designers requires that they be allowed to design. To achieve a good result, they must feel that they can be creative. Listen to them. Try to envision their solutions and suggestions. Rather than stating that something is unacceptable, try saying that you do not understand the solution and cannot agree with a solution you do not understand. You may come to agree with the solution. The designer may develop alternate solutions with which you *do* agree. At the same time, the designer must recognize that you, as facilities or project manager, are in charge and responsible for this new facility. Watch out for the designer that insists that this is his project and that he must be permitted the final word on everything.

Compatibility

As in any work environment, the ability to get along with a working associate is imperative to a quality and efficient job. Assuming that the design firm principals are easy to get along with, ask to meet with the prospective job captain or project manager who will be assigned to the company's project. Some principals will say that they will be your primary contact and that they will attend all of the job meetings. If they *do* attend all of the meetings and do check all of the work as it progresses, either you may get short shrift, or there are no other clients. A design principal who cannot delegate is a one-man operation. Somewhere the work product will not receive the attention it should, with a possible resulting deterioration in quality.

Inquire as to how experienced the designer's job captain or project manager is in dealing with other consultants and, in particular, your suppliers and construction people. The designer will have considerable responsibility in specifying furnishings and construction materials and in reviewing work. However, he will neither be supplying the furnishings nor performing the construction work. Imperious behavior on the part of a designer can cause a lot of bitter feelings among those people who will also be doing the "real" work.

Fees

By in large, most architectural and design firms will be cost competitive in a given area. If one of the firms being interviewed seems too high, you may be paying for reputation. On the other hand, that same high-priced firm may have a superior quality staff which may produce a faster and better product. If the fees seem to be too low, you may be getting low-balled. You will probably pay the difference between the market fee and the low-balled fees in extras. Many firms have job performance standards under which every person working on a project must turn in a minimum number of billable hours every day. If the job is beginning to accrue more hours than the design firm budgeted, you can bet the company will be asked to make up the shortfall or you will begin to see invoices for extras. These extras may or may not be the company's responsibility.

There follows an outline, by phases, which will provide a general description of the usual scope of work performed by a design firm, along with an approximation of the fee allocation, by phase. When negotiating with any of the prospective design firms, have a copy of this outline on hand. If the prospective designer does not provide all of the services described below, it might be wise to consider another design firm. When entering into a contract with a designer, be sure that all of the outlined duties of the designer are spelled out; and that the fee allocation roughly follows the apportionment described in Figure 5.16.

Space Planning Contract Outline

Phase I: Interior Programming

Working with the client, the interior designer will conduct personal interviews and survey all departments to obtain the information necessary for the development of an interior space utilization program, including interpretation of management objectives, space requirements, personal requirements, furnishings, files, special equipment, storage, and departmental adjacency requirements. Working with these surveys, the interior designer will develop the interiors program, stacking and block layouts, and prepare a comprehensive budget for the project, including cost for construction, furniture and furnishings, special facilities, equipment, professional fees, moving costs, and other project costs. The program and budget will be presented to the client for review and approval.

From the approved program, the interior designer will outline special facilities and/or requirements, such as food service, acoustical needs, building security systems, life safety systems, etc., which will require special knowledge, and will secure qualified consultants to conduct surveys in these areas.

Scope Summary:

1. Programming
2. Functional Relationships Diagram
3. Blocking and Stacking Diagrams
4. Client Presentation and Report
5. Consultant Recommendations
6. Special Facilities and Programs

Proposed Fee:

If all seven phases are contracted together, this phase may represent approximately 8.5% of the total fee.

(If the current cost for all seven phases is $2.00/SF, this phase alone would be $.17/SF)

Phase II: Space Planning

Working with the client and with the approved stacking plans, the interior designer will investigate various planning concepts and combinations of concepts for the development of space utilization plans.

Working with approved stacking plans, work space requirements and budgets, the interior designer will prepare concept layouts of typical departments, executive areas, public spaces, etc., showing interior walls and general locations of personnel and equipment, for the client's review and approval.

The interior designer will prepare a budget, establish design and decorating criteria for all interior spaces, and establish the basic criteria for lighting, electrical, telephone, communications and data systems. These criteria shall include:

- Furniture and systems furniture
- Color plates
- Standard and special partitioning
- Floor coverings
- Ceiling and lighting
- Finish items (wall coverings, draperies, accessories)
- Electrical, telephone, communications and data outlet locations
- Window coverings or treatment.

The comprehensive budget will be reviewed and amplified to reflect any adjustments resulting from the concept layouts. This revised budget will be submitted with the concept layouts for the client's review and approval.

Scope Summary:

1. Planning Concepts
2. Workstation Types (if applicable)
3. Preliminary Furniture and Partition Plans
4. Design Concept
5. Preliminary Construction and Furnishings Budgets
6. Presentation

Proposed Fee:

If all seven phases are contracted together, this phase may represent approximately 19.5% of the total fee.

(If the current cost for all seven phases is $2.00/SF, this phase alone would be $.39/SF)

Figure 5.16

Phase III: Design Development

Working with the approved concept documents, the interior designer will prepare preliminary detailed drawings and layouts for final design concepts, including partitions, furniture and equipment, millwork, floor and wall finishes, window coverings or treatments, and interior landscaping.

The interior designer will assist in establishing preliminary selections for furniture and accessory items and will prepare a furniture listing by location.

Budgets will be updated to reflect any necessary adjustments due to the cost of work shown on the preliminary plans, and the cost of furniture, equipment and accessories, all in accordance with the design criteria previously established. Design development and budgeting for special areas such as kitchen and executive areas will be treated with general concept design and allowance budgets for specialty items such as kitchen equipment and custom millwork.

Scope Summary:
1. Design Concepts
2. Floor and Wall Finishes
3. Millwork
4. Plants and Interior Landscaping
5. Furniture Selections
6. Budget Update
7. Presentation

Proposed Fee:
If all seven phases are contracted together, this phase may represent approximately 19.5% of the total fee.

(If the current cost for all seven phases is $2.00/SF, this phase alone would be $.39/SF)

Phase IV: Construction Documents

Upon instructions by the client to continue with the preparation of detail drawings, the interior designer will prepare working drawings and specifications for all areas, which will indicate the location and construction of all items in sufficient detail for the project to be priced and constructed.

The interior designer will finalize and specify, through appropriate and detailed schedules, all architectural finishes, including wall coverings, wood finishes, carpeting, floor coverings, window coverings, fabrics and paint colors, etc., for the client's review and approval. Upon approval, the interior designer will incorporate these finishes into the construction documents.

The interior designer will coordinate and cross-check between necessary engineering and special consultant's drawings or specifications (encompassing structural, electrical, fire suppression, mechanical and special facilities requirements) and the architectural drawings. These documents will be included in the bid or negotiation packages as required.

Scope Summary:
1. Interior Architectural Detailing
2. Cabinet and Furniture Detailing
3. Finish Schedule Preparation
4. Construction Budget Update

Proposed Fee:
If all seven phases are contracted together, this phase may represent approximately 30% of the total fee.

(If the current cost for all seven phases is $2.00/SF, this phase alone would be $.60/SF)

Phase V: Fixture and Furnishings Documents

In addition to the construction drawings described in Phase IV, the interior designer will prepare required specifications documents for interior fixtures and furnishings.

Where new furnishings are required, the interior designer will submit recommendations to the client for approval. Upon approval, the interior designer will complete location drawings showing both new and/or existing items, along with specifications describing all new items in sufficient detail for competitive bidding or negotiation on the particular pieces selected. The interior designer does not purchase and/or handle furniture on a resale basis, but will assist in evaluating and reviewing bids or in negotiations. All purchasing will be in accordance and in conjunction with the client's purchasing procedures. The interior designer will review required shop drawings and approvals for conformance with bid documents.

Figure 5.16 (continued)

Scope Summary:

1. Specifications
2. Scheduling of Delivery and Shipment of Furniture
3. Bid Review
4. Bid Item Approvals
5. Submittal Approvals
6. Keyed Furniture Move-in Plan
7. Furnishings Budget Update

Proposed Fee:

If all seven phases are contracted together, this phase may represent approximately 14% of the total fee.

(If the current cost for all seven phases is $2.00/SF, this phase alone would be $.28/SF)

Phase VI: Construction Administration and Review

Working with the client's selected contractors, the interior designer will make periodic site inspections during the interior construction phase and when the furniture and furnishings are being installed to assure that the work is being done in accordance with drawings and specifications. The interior designer will review the Work progress, and prepare and approve documents for progress payments.

Scope Summary:

1. Site Inspection and Work Review
2. Preparation and Approval of Progress Payment Documents
3. Preparation of Punch Lists
4. Supervision of Furniture Placement
5. Furnishing Inspection and Punch Lists

Proposed Fee:

If all seven phases are contracted together, this phase may represent approximately 5.5% of the total fee.

(If the current cost for all seven phases is $2.00/SF, this phase alone would be $.11/SF)

Phase VII – Follow-up

The interior designer will review and inspect the completed project and approve final progress payments after move-in to assist the client with any miscellaneous revisions or additions.

Scope Summary:

1. Installation Review
2. Approve Final Progress Payments
3. System Fine-tuning
4. Additional Purchases as Required

Proposed Fee:

If all seven phases are contracted together, this phase may represent approximately 3% of the total fee.

(If the current cost for all seven phases is $2.00/SF, this phase alone would be $.06/SF)

Special Services – Executive Areas, Cafeteria, Computer Rooms and Other Complex Areas

Working with the clients, the interior designer will coordinate all services and construction documents related to those special programs which require a high level of special design, including security needs, audiovisual requirements, food preparation areas, computer rooms and other complex areas.

Scope Summary:

1. Space Planning
2. Furniture and Partition Layouts
3. Design Development Drawings
4. Construction Documents
5. Budgeting
6. Bid Specifications and Review
7. Installation Supervision
8. Follow-up as necessary

Proposed Fee:

10%-15% of cost of work specified, including all furniture, fixtures, equipment and construction, less the square footage fee previously noted for these areas.

Figure 5.16 *(continued)*

Refine Budgets and Schedules

Management's approved schematic space plan provides an excellent tool to use in updating the project's Final Budget for approval. Use the plans to begin estimates of such cost elements as the lineal feet of partitioning to be built, the number of light fixtures, electrical outlets, telephone outlets, etc. to be installed. It is time to begin doing accurate carpet area takeoffs, and specifying the numbers of each type and size of furniture needed. Additional information will be produced a bit later in the design development and final design stages. Be sure to obtain the most up-to-date prices for each budget category. Consider current market conditions and use information from local vendors to verify earlier estimates against the actual space plans.

The budget prepared at this point will be the initial stages of the Final Budget for the project. Once it has been reviewed by the Coordinator and approved by the Officer in Charge of the facility, it should not be changed unless program requirements are changed.

As design development tasks are completed, refine the preliminary schedules prepared earlier. If the company has any desire to switch from a manual form of scheduling to a computerized format, this is an appropriate time.

At the completion of space plans and with the submission of the Final Budget for approval, update the schedule, noting any changes which may have come about due to manufacturers' lead times or material and/or labor problems discovered while preparing the Final Budget. If adjustments cannot be made to recover lost time due to delays, in order to avoid paying double rent, notify the Officer in Charge of the facility immediately so that appropriate action can be taken.

Chapter 6

DESIGN DEVELOPMENT

Introduction

Once all the space planning steps described in the previous chapters are completed, it is time to continue the design process of the office space. Though this phase of the project will probably be guided by a professional designer or architect, or the company's own design staff, your views as facility or project manager will be vital to a design that will be functional and pleasing to the company and its employees.

The Design Process

Design affects people every day, from the shape of their morning coffee cup at home, the color scheme of the office in which they work, to the shape and color of their car. Of all the decisions made on a project, the design decisions are the most visible, and among the most important. This book cannot explain how to be a designer, but with the help of experts and with the design elements described in this section, you can participate intimately in the design of a facility. You can decide how and where pre-selected elements are to be used. These decisions shape the final look of the office and the effect the space will have on everyone who works in it.

The company will:

- Decide how to group workstations in the most efficient arrangement, avoiding overcrowding, both in fact and in appearance.
- Determine how major aisles and corridors are to be arranged. For example, it might be beneficial to use a different textured carpet for major traffic aisles, or an accent color on a wall to serve as an orientation point.
- Choose which areas, such as the personnel department, a reception area, or the executive area, may deserve special treatment – perhaps assigning them a different wallcovering or different ceiling or lighting concepts.
- Select areas to receive accent lighting, e.g., elevator lobbies, conference rooms, display walls, training rooms and the like.
- Consider altering the ceiling heights in some areas, such as the lobby or lunchroom, and thus add another dimension to the overall design.
- Select the color scheme that is most pleasing and most appropriate to the planned facility.

These are just some of the factors that determine the final look of the new facility. Because the company will "live" in it for a long time, give thought and care to design decisions. Good design will give great satisfaction year after year.

Detailed Space Plans

A professional from the company staff who is experienced in office design, or the project's designer or architect, will prepare the final plans, incorporating all the basic design considerations described above, the building standard materials, above building standard improvements, new furniture and systems furniture, etc., together with whatever existing furniture and equipment is going to be reused.

The process of developing the detailed space plans begins with the designer creating an inventory of the various components which will comprise the facility. A large portion of this was accomplished in the schematic design process through the creation of the blocking and stacking studies. These studies helped determine the placement of departments, fixed wall facilities and relocatable workstations.

From this base, the designer will begin the process of detailing the precise location of the fixed element components and areas designated and equipped to service any open-plan work areas and stations.

This design process is a three-dimensional one. First, the designer must locate the constructed or fixed elements such as columns, walls, doors, stairways, closets, or chases. Other fixed elements include existing cabinets that will remain or other forms of millwork, etc.

Next will be the location of service elements and devices which must be fed to or are related to the work areas and workstations, such as electrical, telephone and data wall and floor outlets. These must be precisely coordinated with the location of the workstations themselves. It is not sufficient to be close. The location of these service devices must coordinate exactly. If not, they may be in circulation areas, under seating, or possibly directly under a systems panel. Adjustments to accommodate improperly located devices may disrupt the layout of a large area.

Thirdly, all ceiling suspended lighting fixtures must either be coordinated with workstations, other work areas and circulation patterns, or be designed to provide ambient lighting. The designer must be aware of the type of work each workstation occupant is to perform, the amount of light required and its location in relationship to the worker and the workstation's orientation. Once the light fixtures have been located, the designer must coordinate the location of air-handling diffusers and fire suppression sprinkler heads. Actually, fire suppression sprinkler heads are in a fairly fixed pattern, which requires the designer to coordinate the lighting fixtures and air diffusers around the sprinkler heads.

Construction Elements

The detailed space plans will usually include specific information and descriptions of the following:

Walls and Acoustics

Partition Systems Constructed of Gypsum Board on Metal Studs
Partitions usually stop at the ceiling where no acoustical separation is necessary. When sound or acoustical separation is required, the walls may have to be built to reach the structure above the ceiling, filled with an acoustical insulation, and caulked to the structure. The gypsum board may be fire-rated where a fire separation is required, such as a corridor to emergency exit stairs.

Where It Is Used:
Wherever full-height separations between rooms or between office areas and public corridors are required.

Selection Criteria:
Metal studs and gypsum board is the industry standard method for constructing walls, except as modified to satisfy special acoustical needs.

Purchasing:
Normally part of the building standard improvements, which is part of the general contract for construction. Excess partition requirements will be built in an identical manner, but considered above building standard and the excess costs charged to the tenant.

Ordering:
Partitions are not "ordered" per se; consider this item taken care of when the contract is awarded to the general contractor, unless a demountable partition system has been specified.

Installation:
By a partition subcontractor during the early stages of construction.

Special Points:
The most critical item in acoustical separation between rooms is the careful stoppage of sound leaks. Be sure that full-height partitions, meant to be acoustical separations, are thoroughly joined to the floor slab and sealed to the structure above, and carefully sealed where any ducts, pipes or wall-mounted outlets penetrate the partition above or below the ceiling.

Certain jurisdictions require partitions which stop at the suspended ceiling to be braced to the building structure.

Take particular care where a partition meets a window wall. Aluminum and glass are subject to a great deal of thermal expansion and contraction, so the gasket between the end of the partition and any aluminum portion of the window wall must allow for sufficient movement. Use only closed-cell neoprene for such gaskets (not sponge rubber – it will deteriorate in the sun). If linear or "slot" (continuous) diffusers are employed, they may need to be replaced by two diffusers in order to avoid sound transfer, as well as to provide separate temperature control.

Rubber (vinyl) base is normally 4" wide to protect the bottom of the wall from scuffing by cleaning equipment. It should be installed after wall finishes are complete and after the carpeting is installed. The installation and costs associated with installation of rubber or vinyl are usually a part of the carpet contractor's work.

Glass Partitions
Full-height, clearstory, half high, etc. usually tempered, clear glass set in a metal or wood frame which matches the design standard door frames or special framing where appropriate.

Where It Is Used:
On office and conference room walls and other interior rooms. The use of glass partitions or interior windows permits people located in the interior areas of a floor to have visual access to outside views and provides for the harvesting of natural daylight.

Selection Criteria:
Use to provide a feeling of openness, transparency and spaciousness, and to utilize "borrowed" light from windows through one area to another.

Purchasing:
As part of the items purchased through the building standard improvements or as an above standard item.

Ordering:
Again, glass partitions are not "ordered" per se; this item should be included in the general contract.

Installation:
By a partition subcontractor together with a glass and glazing subcontractor during the later stages of construction.

Special Points:
One-inch narrow slat venetian or vertical blinds, or drapes may be purchased to provide privacy when required in interior rooms.

Internal Stairs
An open stairway installed within your leased area.

Where It Is Used:
Usually where there will be continuous traffic between two adjacent floors, and for surges of traffic which would overload the building elevator system. Often used to act as a monumental feature connecting areas such as an executive office suite to a floor containing private dining facilities, special conference rooms or board room.

Selection Criteria:
Normal multi-tenant office buildings' elevator systems are designed to accommodate tenants located on single floors, coming and going throughout the day, without great surges of traffic and without heavy travel between floors. Normal building elevator systems can be overloaded if there is heavy inter-floor travel or personnel movement to lunchrooms, cafeterias, etc. An inter-floor stair can often improve the effectiveness of the elevator system and prevent long lines waiting for elevators. Be very cautious about putting an internal stairway directly into a lunchroom area. Those enticing aromas coming through the open stairway during lunch may not wear too well at other times of the day. Also, be aware that creating a floor penetration, and constructing such a stairway is very expensive. In addition, this kind of floor penetration may have code consequences. Be sure to check with the project architect.

Purchasing:
As part of the items purchased as an above standard item. This is normally an above building standard element, the cost of which will be charged to the tenant.

Ordering:
Should be part of the contract awarded to the general contractor.

Installation:
By the various trades employed by the general contractor during the normal construction cycle.

Special Points:
Since an inter-floor stair occupies floor area for which the company will be paying rent and which could otherwise be used for office occupancy, carefully review the cost-effectiveness of installing one. The personnel time lost by people waiting for elevators can often quickly pay for the extra rent and cost of the inter-floor stair. Be sure to review the anticipated daily activity in the proposed facility with the designer or architect during preparation of stacking plans, in order to correctly determine the need for an inter-floor stair.

Doors, Hardware, Locks and Keys

Doors, doorknobs, locks and other hardware required for entrances, private offices, conference rooms, and storage rooms should be selected to satisfy all of the company's security requirements. Door closers, special hinges and other devices may be required by building codes.

Where They Are Used:

Throughout the office spaces as required.

Selection Criteria:

The landlord usually has selected a building standard hardware and is not often inclined to deviate from his selection. This is particularly true of entry door hardware. It may be possible to persuade the landlord to permit an alternate hardware in interior spaces. Keep in mind the implications of ADA requirements.

Purchasing:

Normally part of the building standard improvements as part of the general contract for construction. Using an alternate hardware may incur an above building standard cost.

Ordering:

Should be part of the contract awarded to the general contractor.

Installation:

By the general contractor during the late stages of construction.

Special Points:

Generally, a 3' x 7' solid-core door, with plastic laminate or wood veneer face will be building standard. Hollow metal doors are more durable, but are more expensive and do not look as attractive as wood doors; only use them where a door is expected to take a beating. Hollow-core wood doors are generally too lightweight for commercial use.

Be sure to discuss the company's specific security requirements with the interior designer or architect for recommendations on which areas should be locked, levels of keying, and other technical data.

If the company is leasing and the building standard doors or hardware appear unsatisfactory, request technical data on them from building management and ask your interior designer to review them. (The building standard door frames, hardware, locks and butts should be reviewed to assure that they comply with minimum standards.)

Ceiling Systems

The most common ceiling is the exposed grid or "T" bar suspension system with lay-in ceiling tiles or panels. Try to achieve a noise reduction coefficient (NRC) of 0.60 to 0.75 or better. Most surface finishes are fissured or matte. The industry offers a very wide selection of grid suspension systems and panels. A designer or architect can achieve some very interesting designs, while possibly improving light reflectivity and sound control. For example, as an inexpensive alternative to the very frequent 2' x 4' system, you might specify a 2' x 2' grid with tegular panels and a glacial-type finish for a more stylish appearance. Most buildings provide 2' x 4' as a standard, but the landlord may be willing to convert to a 2' x 2' system, which will provide greater flexibility, for a moderate extra charge. Upgrading the ceilings need not be universally applied. Selecting different systems and panels in different or critical areas can not only be aesthetically pleasing, but can help to enhance acoustical and lighting control. Some rooms, such as executive areas, may be finished or partially finished with gyp-board type ceilings.

Where It Is Used:
Throughout all general office areas.

Selection Criteria:
Fiberglass usually has a better ability to absorb sound than the mineral fiber panels normally used in office buildings. However, the surfaces are usually either a thin layer of vinyl or cloth fabric. The vinyl is inexpensive and not too attractive. The cloth fabric is very attractive, but relatively costly. The more common selection is mineral fiber board. It comes in a wide variety of thicknesses, NRC and STC ratings, surfaces, colors and sizes. When installing ceilings in open-plan areas, keep in mind the probable need for flexibility and the possible desire to relocate lighting fixtures and air-handling diffusers.

Purchasing:
Normally part of the building standard improvements as part of the general contract for construction.

Ordering:
Should be part of the contract awarded to the general contractor.

Installation:
Hanger wires are attached to the structure to suspend the grid system. Panels are laid into this grid after the installation of above-ceiling duct, sprinkler, telephone, data communications and electrical work.

Special Points:
Since the fiberglass ceiling will not stop sound effectively between private offices where partitions stop at the ceiling, those partitions separating private offices from other offices, conference rooms, and other work areas may need to be extended above the ceiling to the structural slab and sealed against sound leaks. Ask the interior designer or architect for assistance in determining where acoustical partitions should be installed.

In large open-plan areas, certain building codes may require that the plenum (the area above the ceiling) be broken down into smaller zones by means of walls constructed from the ceiling to the structure above.

In certain seismic zones, ceilings must be laterally braced to keep panels from falling out in the event of an earthquake. The suspension wires must also be placed within specific distances of each corner of all lighting fixtures held by the ceiling grid.

Check with the project architect or local code authorities to be sure planned improvements comply with all local codes.

If the ceiling system offered by the building is different from what the company desires, discuss the implications of the different system on the acoustics of the new space and request recommendations from the architect or interior designer.

Millwork or Cabinetry

All millwork should be premium grade, with all exposed surfaces clad with wood veneer or plastic laminate. Consider the use of free-standing metal shelving wherever possible, for lower cost and higher structural capacity. It can often be used in storerooms which are not exposed to public view.

Where It Is Used:
Lunchrooms, work rooms, mail rooms, copy rooms, supply areas, and miscellaneous areas where built-in cabinetry is needed or desirable. Review all millwork and cabinetry locations with the designer.

Selection Criteria:
This is the industry standard method of accommodating requirements in these areas. (Many companies may also use free-standing metal cabinets in some back house areas, i.e. copy/supply/mail.)

Purchasing:
Normally in addition to the building standard improvements as part of the general contract for construction.

Ordering:
Should be part of the contract awarded to the general contractor.

Installation:
By a millwork or cabinetry subcontractor during the later stages of construction.

Special Points:
Require that shop drawings be provided by the millwork contractor who will fabricate and install the work. These shop drawings should be submitted to the interior designer or architect for review and approval.

Where large amounts of storage have been designated, be sure to check the allowable floor loading with the landlord; request that he tell you the allowable weight per square foot of floor space or shelf unit, and minimum aisle dimensions between shelf units.

Lighting and Switching
Fluorescent lighting fixtures are usually part of the building standard improvements. Normally 2' x 4' (although it may be 1' x 4' or 2' x 2'), with switching to allow fixtures to be switched off in any room or area, and to allow half the tubes in each fixture (alternate fixtures where 2-tube fixtures are used) and perimeter rows of fixtures to be shut off when full lighting levels are not required. In addition to the general or ambient lighting, the lighting engineer or architect may employ incandescent down lights or "wall washers" for visual interest or to emphasize a particular point of interest.

Where It Is Used:
Throughout office areas. Because of the heat generated by incandescent lights, they should never be used directly over a point where someone is working or sitting.

Selection Criteria:
This is the industry standard and normally part of the building standard improvements.

Purchasing:
As part of the building standard improvements, which are part of the general contract for construction. The switching described above may not be provided by the building in areas which do not have energy conservation concerns, but should be specified as part of an energy conservation program.

Ordering:
Should be part of the contract awarded to the general contractor.

Installation:
As described above by an electrical subcontractor, after the ceiling suspension system is in place.

Special Points:

Most general office spaces are lighted with cool white or warm white lamps, since most wall and fabric color selections will look best under these colors of lamps. Deluxe cool white lamps will have a Kelvin temperature of about 4100K and a CRI (Color Rendering Index) of about 89. Deluxe warm white will be about 3000K with a CRI of about 77. The ideal color rendering is in the 4000K range with a perfect CRI of 100. Please see Chapter 16 on Lighting for more information.

A limited number of recessed incandescent down lights may be used to highlight art work or graphics in reception areas or lobbies.

Specify "low brightness" type lenses wherever available to reduce glare in office areas. Contact the interior designer for help.

Be sure that fixtures are installed in compliance with all local codes, particularly in earthquake-prone areas.

Always have an expert review the proposed lighting plan or the one provided by the building in order to assure proper light levels and distribution. Furniture-integrated lighting is an available alternative. This is intended to correlate the lighting fixtures with the furniture system, thus reducing or eliminating the need for the conventional overhead ambient system. This alternative may improve lighting quality and acoustics, and reduce energy consumption.

Electrical, Telephone and Data

The electrical and communications systems distribute electrical power, telephone and data cables throughout the office space to wall and floor outlets for systems furniture and equipment.

Selection Criteria:

This is an industry standard and normally part of building standard improvements. Special outlets may be necessary for connection to the systems furniture which are part of the standards or for computer terminals or networks.

Purchasing:

Part of the building standard improvements, which are part of the general contract for construction. Special outlets required for systems furniture connections or computer networks may normally be "traded" for electrical and telephone outlets which are provided as part of building standard improvements, and may often result in a credit to the tenant. This is a complex matter which will require the assistance of the architect, interior designer or electrical engineer. The tenant will also normally be required to pay for special voltages, single dedicated or clean circuits and/or isolated grounds required for some telephone equipment, computers, special data entry terminals, copiers, vending machines, etc., which are normally not included as part of building standard improvements.

Ordering:

Should be part of the contract awarded to the general contractor.

Installation:

Conventional duplex electrical outlets and telephone outlets are mounted on the walls in private offices, conference rooms, and work rooms. Special junction boxes are installed in the walls or floor to feed power, telephone and computer cables into the base of systems furniture, where used.

Outlets must be provided wherever electrical appliances, telephones or computer terminals are to be used, as well as for cleaning purposes.

Special Points:

Be sure to carefully check and double check the dimensions for the location of outlets which will serve systems furniture. Check the electrical subcontractor's marks on the floor before he drills holes to install the outlets. The location of the floor outlets is very critical to the location of systems furniture.

Be sure to carefully review each electrical and telephone outlet shown on the plans against the company's specific office requirements to be sure all the company's needs are accounted for. Do not assume that the person preparing the drawings has provided for all your needs or located the outlets in the correct places. Learn to read the construction documents; do not delegate this task.

Extension cords are never to be used for power to appliances. Always install an additional outlet when the length of the cord attached to the appliance is exceeded.

Lighting Design

Location of the building standard light fixtures and any special light fixtures is made at this point in the design process. The professional preparing the documents is likely to prepare more than one lighting layout sketch that takes into account the following criteria:

- The coordination of heating, cooling and ventilating elements and diffusers, and fire suppression sprinkler heads with soffits and partitions.
- Conformity to the building standard ceiling grid, if any;
- Location of workstations and conference tables to avoid glare and veiling on work surfaces or VDT screens;
- Special lighting requirements for artwork and graphics, or signage and display areas;
- Energy conservation through double switching and zone lighting. (For example, if light fixtures next to windows are placed on a separate switch, they can be turned off when daylight is adequate; or two of the four lamps in a normal 2 x 4 fixture can be turned off when not required because of adequate daylight or for daily housekeeping, thereby conserving energy).

Music and Paging

An amplifier and speaker system providing pre-recorded music and the ability to page personnel may be desirable or even a requirement.

Where It Is Used:

Throughout the office space except in private offices and conference rooms.

Consider piping music to public lobbies, corridors and toilet rooms on floors fully occupied by the company, as well as to work areas. Paging may be selected by area, with the equipment and paging microphone located at the security office, main switchboard, human services office or the office manager's office.

Selection Criteria:

Music and *random* or *white noise* systems help mask general office sounds, creating a more pleasant work atmosphere; however, be careful in choice of musical format and frequency of play, as many employees have a strong negative reaction to the "elevator music" sound. Paging ability should be installed if you find it necessary to locate people quickly when they are away from their workstations. Many acoustical engineers will likely suggest *random noise.*

Ordering and Installation:

Music and paging systems are generally, although not always, purchased on a *design-install* basis through local and national vendors. Discuss your requirements with the vendor the company has selected; the vendor will then prepare design documents, secure permits, coordinate its work with the general contractor, and provide a fully operating system for the facility. The system should be ordered at the completion of construction documents and installed at the same time as the ceiling and lighting.

Special Points:

On a large installation, such as a regional office or headquarters building, secure professional consultation, as opposed to purchasing on a design-install basis.

If speakers are located in fiberglass-lined metal boxes above the ceiling, and pointed up toward the structure so that music will bounce off the structure and filter uniformly down through the ceiling panels, no penetration of the ceiling panel needs to be made. Fiberglass ceiling panels are acoustically transparent for music and paging purposes, and for this purpose may be desirable over other types.

Speakers mounted in the ceiling tile and directed downward may create objectionably loud areas directly below the speakers if music and paging are to be loud enough to be heard in all areas. Make particular note of this point when discussing installations with vendors.

Some local jurisdictions will require fire department or other life safety paging systems throughout the building. Some buildings provide such systems. Be sure to make every attempt to coordinate the installation of the system so that it ties in with any life safety system within the building.

The amplifiers and all equipment should be located within a secured area so that only those authorized to do so may make adjustments to the system.

Review the pre-programmed music systems available from local vendors to determine which program of music best satisfies the requirements of your facility and the tastes and desires of your personnel.

Furnishings and Finishes

Carpet

Two types of carpet are in general use:

- Cut-pile, tufted carpet made of Antron IV nylon or other similar fibers. Cut-pile carpet has a plusher look and feel, but it does not wear as well as loop pile carpeting. It tends to *shade*, that is the pile lays over in higher traffic areas and looks as if it is wet or *shaded*.
- Loop pile, tufted or woven construction carpet, made of Antron IV nylon or other similar fibers. Loop-pile carpet wears longer and better than cut-pile carpet. However, it does not have the rich look of cut-pile carpet.

Where It Is Used:

Cut-pile carpet is used in reception areas, private offices, conference rooms, and other high profile and/or low-traffic areas. Loop-pile carpet is used in operational areas, corridors, and similar high-traffic areas. The use of carpet tiles is popular with some companies. There is much to be said for its adaptability to change and repair, especially if a textured design is selected. Wood flooring and some natural stone may be used under desk chairs, in lobbies or as a carpet border. Vinyl tile may be used in special areas such as copy centers, or vending machine areas, etc. See Chapter 11 on Floor Coverings for more information.

Selection Criteria:

Most companies have varied carpet requirements, most of which are satisfied by the two carpet types just described. The most critical carpet wear problem in any office facility is that of desk chairs rolling on the carpet. A woven carpet should be considered in areas where personnel are frequently moving to and from their workstations. It also wears the best. Corridors and private offices get much less abuse from the rolling of desk chairs; therefore, a carpet made of a tufted construction is adequate for these areas.

Purchasing:

Most carpets come in rolls 12 feet wide by approximately 100 feet long. Carpet prices are quoted by the square yard and a separate price per yard is quoted for installation.

Ordering:

Woven carpet requires long production times, so allow approximately 16 weeks after placing an order for its delivery. Tufted carpet can be supplied in ten to twelve weeks, or less.

Installation:

Glued directly to the floor slab. Specifications can be found in the Master Forms Section.

Wall Finishes

Three types of wall finishes are likely to be used in typical facilities. They are paint, vinyl, and fabric. Wood paneling or stone will occasionally be used in executive or reception areas, and ceramic tile, laminate and stainless steel will be employed in speciality areas such as food preparation and serving areas.

Where It Is Used:

Paint is used on the majority of wall surfaces in operational work areas. Vinyl wall covering is used in areas subject to heavy traffic, (such as corridors, lunchrooms, and elevator lobbies.) Fabric may be used in reception areas, conference rooms and private offices.

Selection Criteria:

- Paint is the more economical wall finish and is easiest to apply.
- Vinyl wall covering is slightly more costly, but often the added expense is justified where maintenance is a problem, such as along corridor walls or wall corners at circulation points.
- Fabric is the most costly; however, it is also the most aesthetically pleasing of the three. It should be used only in special areas, at the direction of the person designing the project and with the approval of the Coordinator.

Purchasing:

Paint is typically part of the building standard improvements and included in the general contract for construction. Vinyl and fabric wall coverings may not be included in the landlord's building standard improvements, and will have to be included as an above standard item and the excess costs charged to the tenant.

Ordering:

Finishes are noted in the construction drawings on the Finish Plan, and will be included in the general construction contract.

Installation:

By the painting and wall covering contractor under the direction of the general contractor.

Window Coverings

Draperies, narrow horizontal blinds, Mecho-type shades or vertical blinds for all exterior glass areas.

Where It Is Used:

On all exterior windows as required for solar heat or lighting control or when desired for aesthetic reasons.

Selection Criteria:

Horizontal blinds are often preferred because of their ability to control sun infiltration and still allow a view to the outside; draperies are used where a softer look is desired. The choice between vertical or horizontal blinds is often an aesthetic one.

Purchasing:

Usually as part of the building standard improvement. However, they may sometimes be a tenant improvement. If the building is one without solar-control glass, blinds will be needed to assist in maintaining comfortable temperatures. Under these conditions, blinds should be considered as a building standard. Even with solar-control glass, you may wish to use blinds to reduce glare, especially on VDT screens.

Ordering:

Should be covered under the general contractor's work.

Installation:

Usually by a drapery subcontractor at the completion of construction.

Special Points:

Contact your interiors designer for assistance in reviewing the building standards or in specifying alternatives.

If you use horizontal blinds, provide a ring pull located at the head rail of the blinds to raise and lower them for cleaning purposes. Blinds should not be capable of being partially raised. Pitch adjustment should be left to individual control. The lack of continuity of pitch may not be aesthetically pleasing, but sun and glare control may be a necessity.

Design Details

Once the designer has completed the detailed space plans, a set of *Detail Drawings* will have to be prepared. These drawings explain how specific elements are to be constructed or fabricated and how such things as cabinets are to look in plan, elevation and sectional views. The Detail Drawings may include most of the following:

- Special walls (e.g., fire rated, security and acoustical partitions);
- Special lighting (e.g., cove lighting in elevator lobbies or recessed lighting in conference rooms);
- Special ceiling materials (e.g., gypsum board ceilings in such areas as executive reception areas, or testing labs);
- Special Millwork and Cabinet Drawings and Details
  Special millwork includes items more akin to furniture than construction elements, such as bookshelves and conference room credenzas, as well as built-in storage closets, mail room sorting counters and computer tape storage areas. These drawings should be done by the person designing the project in sketch form for review. Once approved, these sketches may be completed as finished drawings and incorporated into the construction drawings detail sheets, or they may become special drawings to be sent out to a millwork or cabinet shop or supplier as part of the furniture procurement process.

Furniture Selection and Location

At this point in the design development process, the company must make a final module selection and standards selection. The further development of the drawings depends on a specific set of furniture and systems furniture dimensions. It is not necessary to have selected a specific manufacturer. However, you should have selected at least two furniture lines that will satisfy your standards. The designer will begin the process of producing the Furniture Drawings and placing specific workstations on the drawings. Each workstation will have its component or furniture configuration displayed. Each piece of furniture, new or relocated, should be shown and identified on the Furniture Drawings. Your coordinator should work very closely with the designer during this phase of the drawing preparation in order to assure that what the designer is drawing represents what each workstation is to be like when the actual space is built and all of the furniture is put in place. These drawings will also be used in the furniture procurement process.

Finish Selection and Locations
(Walls, Floors and Ceilings)

The development of these drawings will involve deciding how color and materials will be used throughout the project. It may be wise not to establish rigid color and design standards for paint and wall finishes. Cultural and environmental tastes vary widely from one section of the country to another and from one facility to another. The professional who designs the new project should meet with the Manager or Officer in Charge of the facility and review the color scheme, and carpet and upholstery options. Once a decision is made as to a color theme, the designer will prepare a finish plan for review. This plan will show the location of the various carpet selections, all wall, door, millwork, counters, frame and trim finishes, ceiling panel types, and any special floorings (e.g., wood, vinyl composition tile, etc.).

Final Design Elements

Color and Fabrics

Color and fabric decisions will be made for upholstered and free-standing furniture and for the two or three alternate systems furniture panels. The location of each item of furniture will be noted by the designer for transfer to the final furniture plan.

Plant Selection and Location

This is usually one of the last items to be completed on a project. Consult with several local nurseries to determine which plants are best suited to your specific facility. After approval, produce the plant location plan. Please see Chapter 14 on Plants and Containers.

Artwork Selection and Location

Art is the finishing phase of a project. The Coordinator or art consultant should select several alternatives for each specified location within the facility. After final selection of the art, the Coordinator or consultant will complete the framing and prepare for installation.

Revised Budget and Schedules

At the completion of the design development phase, it is time to do yet another revision to the budgets and schedules. During this phase, make constant revisions to the budgets, especially to the detail budgets which will become the back-up to the principal line items of the budget. If it appears budgeted costs will be exceeded, there are several alternatives:

1. Proceed with the current design direction and hope to cure the over-costs through the construction bidding or negotiation process. If the over-costs can't be corrected as a result of the bidding process, go back to the design development level, re-do those and subsequent construction documents and re-bid the project. This process will carry additional design costs and probably delay the job by a minimum of 60 days; or

2. Ask the Officer in Charge for more money to cover the estimated over-cost; or

3. Try to attack the various cost elements by eliminating certain items or by having them priced as add-alternates that can be included only if other costs are lower and the budget will permit. Consider moving to lesser quality components and finish and/or attempt to re-allocate potential budget savings from one line item to another.

It may be that in spite of all best efforts, you are unable to stay within the original budget parameters. In this case, there is no option but to eliminate certain items or to ask the Officer in Charge for a budget increase.

Chapter 7

CONSTRUCTION DOCUMENTS

Introduction

Before proceeding with the steps discussed in this section, be sure to try to submit the preliminary space plans to local building authorities for a preliminary check. (Be aware they may not have the time or manpower to do preliminary plan review.) Try to have made any requested revisions or corrections, and have received approval from senior management to continue with the project.

Construction Responsibility

At this point in the development of the project, either the tenant company's agent or the landlord's architect will proceed with the preparation of construction documents. The decision as to who will prepare the construction documents should have been decided at the time the Lease and Tenant Work Letter were agreed upon and prepared. (See the Chapter "Evaluating Space Options" for further information.) If the tenant was to be permitted to do the actual construction, then it follows that they will prepare the construction documents. These documents will undoubtedly have to be submitted to the landlord for approval before any work will be done. However, if the landlord is responsible for the construction, it is most likely that his architect will prepare the construction documents, which will be based upon space plans prepared by you as facility or project manager.

At this point, review the portions of this handbook which deal with the preparation of the Tenant Work Letter and the Lease. See Chapter 2 for further information. The process by which the tenant and/or landlord controls the construction should have been worked out in detail during the lease negotiation process. If not, review that section in order to know how to proceed at this point in the project process. Also review the lease as a reminder of the company's rights and obligations. The likelihood is that the tenant will have to arrange for an agreeable payment process for above standard work. If the landlord is doing the work, the tenant will wish to assure himself that the work being performed by the landlord's contractor meets his specifications and is performed in an accurate and timely manner. While monitoring the landlord's contractor's work, do exercise extreme care in dealing with the contractor's personnel. Remember the tenant and his representatives have no privity of contract with the contractor. If some form of clarification is needed, go through the landlord's representative.

If the tenant or his agent is to manage the construction process, he will need to have prepared additional drawings and specifications which define the scope of work. These drawings and specifications are known as construction documents.

Preparing Construction Drawings

Before the design concept can be implemented and the space actually built, the work must first be described in what are known in the construction industry as *construction documents* or *working drawings* which describe in detail the work to be performed. These documents consist of a series of drawings primarily describing any demolition, construction, construction details, finishes, and often the furnishings layout drawings. Accompanying the drawings is a package of specifications and special conditions which will describe the specific materials to be used in the project, the acceptable manufacturers, and the conditions and techniques for installation or application. These documents are used both to obtain bids from contractors who will do the actual construction, and also to obtain necessary permits and approvals from the necessary authorities. They should be prepared by a professional. The law in many jurisdictions may require that the documents be prepared and checked by a licensed architect.

Building Standards Information

In order to proceed, have your landlord or building manager fill out the form Building Standards Information. A sample is illustrated in Figure 7.1. It is also found in the Master Forms Section. It would be beneficial to ask the landlord to fill out the Building Standards Information form before the lease is signed. Your landlord will be much more cooperative before the lease is signed than after.

Drawing Formats

If the company has multiple facilities, the person preparing the documents for this Project should follow a standard format in order to maintain a consistent approach throughout all company facilities. This approach should be a step-by-step process designed to provide complete information on each drawing and to provide a consistent level of information for this or other projects.

Facility: _____ Location: _____

Prepared By: _____ Phone: _____ Date: _____

Owner: _____ Contact: _____

Area: Rentable: _____ Usable: _____ Plannable: _____

Complete this questionnaire before beginning the design process. Be as complete as possible. One interview with the Landlord's Building Manager or representative should be able to secure most of the information. If data is not available, so indicate. If not applicable, so indicate. Try to respond to each question. You will save yourself considerable backtracking.

Address: _____ Floor(s): _____

City: _____ State: _____ Zip: _____

Representative: _____ Phone: _____

Which building code(s) govern?

1. **Building architect:**

   Address: _____

   City: _____ State: _____ Zip: _____

   Representative: _____ Phone: _____

2. **Engineering firms:**

   Electrical: _____

   Address: _____

   City: _____ State: _____ Zip: _____

   Representative: _____ Phone: _____

   Mechanical: _____

   Address: _____

   City: _____ State: _____ Zip: _____

   Representative: _____ Phone: _____

Figure 7.1

Plumbing: _____

    Address: _____

    City: _____ State: _____ Zip: _____

    Representative: _____ Phone: _____

Structural: _____

    Address: _____

    City: _____ State: _____ Zip: _____

    Representative: _____ Phone: _____

3.   Issuing of plans: When drawings are complete, to whom should they be issued? How many to each?

    For building permits: _____ Quantity: _____

        Address: _____

        City: _____ State: _____ Zip: _____

    For construction: _____ Quantity: _____

        Address: _____

        City: _____ State: _____ Zip: _____

    For other approvals: _____ Quantity: _____

        Address: _____

        City: _____ State: _____ Zip: _____

4.   Partitions - type (description) : _____

    Demising walls: _____

    Public corridor partitions: _____

    Standard interior: _____

    Sound attenuation: _____

Figure 7.1 (continued)

5. Building standard entry door & frame (sketch w/dimensions or descriptions):

_____

_____

_____

Knockdown or welded (circle one) - Aluminum or steel (circle one)

6. Building standard interior door & frame (attach sketch w/dimensions or descriptions):

_____

_____

_____

Knockdown or welded (circle one) - Aluminum or steel (circle one)

7. Building standard hardware: _____

   Model: _____ No.: _____ Finish: _____

8. Floor base: _____ Height: _____ Color: _____

9. Building standard floor covering: _____

   Pattern No.: _____ Size: _____ Colors: _____

10. Building Standard paint: _____

    Series or No.: _____ Number coats: ____ Plus primer?: ____

11. Building standard induction unit finish (if applicable): _____

12. Building signage restrictions (describe): _____

_____

_____

_____

_____

_____

_____

Figure 7.1 (*continued*)

13. Building standard window covering: _____

Slat width: _____ Series No. _____ Color: __

(Attach sketch of head, sill and jamb details, drapery track, linear air diffuser with dimensions.)

14. Building standard electric water coolers: _____

Model No.: _____ Finish: _____

15. Public corridor: width: _____ Floor finish: _____

Wall finish: _____

16. Floor structural capability - general office areas: _____

Dead load: _____ Live load: _____

17. Building standard emergency speaker locations: _____

18. Ceiling specifications: _____

Tile size: _____ NRC: _____ STC: _____ Series: _____ Color: _____

Suspension system description: _____

19. Sprinklers?: _____ Wet or dry system?: _____

Building standard sprinkler head: _____ Pendant: _____

Unspoiler: _____ Flush mtd. _____ Finish: _____

20. Ceiling breaks for major ductwork: _____

Finish floor to bottom of structure: _____

Ceiling height: (Attach sketch) _____

21. Return air system: Plenum: _____ Ducted: _____

22. Typical thermostat locations: _____ Description: _____

23. Heat from lights used for building heating?: _____

24. Number of fixtures per square foot provided: _____

25. Building standard fluorescent light fixtures: _____

Size: _____

Figure 7.1 (*continued*)

Manufacturer: _____ Model: _____

Lamps description: _____ Lens description: _____

Ballast description: _____

27. Building standard electrical switches: Ht.: _____ Type: _____ Color: _____

28. Building standard electrical outlets: Ht.: _____ Type: _____ Color: _____

29. Description of building standard electrical switching system: _____

_____

_____

_____

_____

30. Can conduit for outlets, switches, or telephones be installed within existing:

   Core walls: _____ Demising walls: _____ Columns: _____

   Perimeter walls: _____ Perimeter columns: _____

31. Under-floor electrical raceway system (description): _____

   No. cells: _____ Cell sizes: _____ L.F. on center: _____

32. Exit signs: _____ Model No.: _____

   Color of letters: _____

33. H.V.A.C. (description of system): _____

_____

_____

   Diffuser types: _____

   Sizes available: _____ Color: _____

34. Elevator doors-finish: _____ Color: _____

   Location and description of call buttons and lanterns: _____

Figure 7.1 (*continued*)

35.   **Miscellaneous information:** _____

_____

_____

_____

_____

_____

_____

_____

_____

_____

_____

_____

_____

_____

_____

_____

_____

_____

_____

_____

_____

_____

_____

_____

Figure 7.1 (*continued*)

CSI MasterFormat – We have prepared a sample set of construction drawings (documents) for the third floor of the sample space described throughout the book. The assembly of the sheets will basically follow the Construction Specifications Institute's (CSI) MasterFormat. This system is used by most architects, engineers, contractors and manufacturers for classifying work and products as shown in Figure 7.2.

There is a sample set of construction and furnishing drawings at the rear of the Master Forms Section. It will be helpful to examine them before continuing with this Chapter. The drawing set will closely follow the CSI format, modified to follow a logical construction, finish and furnishing progression.

Note that the CSI index has Furnishing 09, ahead of Electrical 16. Our drawings place Electrical 3-2, ahead of Furniture & Furnishings, 3-F2, because this is the order in which the work, furnishings, bidding and placement will progress.

Sample construction and furnishing drawings should be supplied to the professional who will prepare a set of similar drawings for the proposed facility.

Most construction drawings will be produced on 30" x 42" drawing paper. The *blue-line prints* will be exact reproductions of the original *hard line drawings*, and will also be 30" x 42". There may be occasions when the architect or designer may prefer to employ a different page size because area dimensions would be better suited to another size. We have placed the *title block* along the right hand edge so that the sheet title and number can be quickly located in a large bundle of drawings. You should ask the architect to also set up the *title block* on the right hand edge rather than across the bottom of the drawing.

CSI Number	Description
01	General Requirements
02	Site Work
03	Concrete
04	Masonry
05	Metals
06	Wood and Plastics
07	Thermal and Moisture Protection
08	Doors and Windows
09	Finishes
10	Specialties
11	Equipment
12	Furnishings
13	Special Construction
14	Conveying Systems
15	Mechanical
16	Electrical

Figure 7.2

As with any of the samples contained in this book, do adopt a format best suited to the company and its style of developing its facilities. The important consideration is to try to establish a pattern of consistency and manageability. This becomes more important, the more active space management becomes.

Construction and Furnishing Documents

To familiarize you with a typical drawing set and the production approach, there follows a description of each basic drawing and its title or alternate titles, and a typical page numbering system.

Title Sheet Blank Title, No Number

A blank sheet with your logo, followed by the name of the architect or designer who prepared all the drawings. This title sheet is prepared once and then is reproduced photographically on each of the subsequent drawings. After this is done, each of the drawings receives its respective title.

Core Drawing (Or Background Drawing) Blank Title, No Number

This drawing is not specifically a Construction Drawing but is a necessary step in the development of the Construction Drawings. If your work is being drawn on a CAD or CADD system, this drawing is the first one prepared in anticipation of all of the others. It consists of an accurate reproduction of all the structural and permanent existing walls and spaces of the building, including existing toilet rooms, stairways, columns, elevator shafts, telephone and electrical closets, air shafts and the like (darkened on the drawing to distinguish them from any new construction).

If hand prepared, this completed drawing is then reproduced to produce drawings 3-1, 3-2, 3-3, 3-4, 3-F1, 3-F2, and 3-PL (described below). It is important that the core information appear on each of these drawings. This drawing may also establish a grid nomenclature designating column lines with letters reading from left to right and numbers reading from top to bottom. Use of this system will let you locate and describe an area on a floor for example as "...being to the right of and below column C-3". Whichever way the plans are drawn, be sure that they reflect the *as-built conditions*. The Base Building drawing should be field checked to be sure that it accurately represents the building and *as built* conditions. Failure to do this could lead to disastrous surprises during construction.

Master Legend, ML-1

Contains and explains the symbols, codes, and specifications used in the other drawings for such items as partitions; lighting, electrical and telephone systems; wall coverings and carpet. In addition, it serves as a checklist to ensure that every piece of information required to complete the project is transferred to the contractor(s) by way of drawings and/or specifications. Quite often this information appears on the individual drawings and in the specific sections of the specifications for the respective trades, rather than on a single combined sheet.

Construction Plan, 3-1

Sometimes called Partition Plan. Sometimes designated with an "A" for Architectural, such as A-3. This plan (or plans) will show the various types of partitions, (drywall, special sound wall, or glass integrated with drywall), their locations, and distance from existing core areas or columns. There will also be heights, door symbols, any special construction details, and elevation reference symbols. The construction plan serves as the main reference document for cross-reference of coded symbols and details. The coded symbols tell you on which drawings you will find additional information. Before coded detail symbols are added to the Construction Plan, it is also reproduced for preparation of drawings 3-2, 3-3, 3-4, 3-F1, 3-F2, and 3-PL. Those drawings should show all partitions.

Telephone and Electrical Plan(s), 3-2

Will also include Communication and Data Information. Usually prepared by appropriate engineers. They will show the telephone, electrical, communication, and data installation contractors the proper locations for the equipment and complete information about it. These plans also indicate whatever special separate circuits and special voltage outlets the facility may require. These plans will have schedules showing the required types of outlets, voltages, wattages, etc. for the various outlets.

When using systems furniture with electrical and telephone wiring capabilities in the panel chases, this electrical information must be given to the various contractors and engineers prior to the preparation of their construction documents.

Reflected Ceiling Plan, 3-3

This plan or plans will show the type and height of the ceiling, the location of the grid for the ceiling suspension system, the direction of ceiling tile fissures, and the precise location of light fixtures, diffusers, speakers, sprinkler heads, soffits, etc. It may contain special and/or construction details, along with section marks, special ceiling finishes, etc.

Finish Plan, 3-4

Shows the locations of the various types of floor coverings (carpet, vinyl composition tile, wood, ceramic tile, etc.). You can use the plan to calculate the square footage of these coverings. In addition, the Finish Plan indicates the locations of any specified wall finishes such as paint (colors), vinyl, fabric or acoustical materials.

Details, D-1

Drawings that clarify how the variety of building materials fit together. For example, how the drywall partition meets a metal door frame, or relates to a ceiling system.

Elevations, D-3

Drawings that show vertical conditions that are being constructed or coordinated with existing conditions or with other trades. They are used to further define such items as ceiling, door and glass heights, custom cabinet work, and any special drywall or millwork construction that cannot be clearly explained on the Construction Plan. This plan may also indicate finishes not easily shown on the Finish Plan.

Panel Plan, 3-F1

Necessary only when systems furniture is being used. It shows the exact location, number, and code of each panel, and all of the workstation systems components cross-referenced to the panel specifications. The same numbering system should be followed for all plans so that all furniture specifications, architectural drawings, and furniture plans use the same system. Unless the general contractor is installing the systems furniture he may not need a Panel Plan. It may be preferable to supply the general contractor a "For Your Information" copy especially for electrical coordination.

This plan must be very carefully coordinated with the electrical plan and with building structural conditions for the placement of floor outlet boxes. These boxes consist of either a service fitting head tapped into an under-floor duct or into a poke-through. Floor outlet boxes (floor monuments, "dog houses", etc.) are quite costly to install and often can not be placed exactly where desired because of the modularity of the under-floor duct system or because of conflicts with the structural or mechanical systems or other interfering elements. As previously mentioned, the exact location of these boxes should be field checked prior to tapping into the under-floor system or performing the core drilling in order to be sure that they will coordinate exactly with the furniture systems panel locations. It is highly recommended that the company's architect assist in coordinating the core drilling in the field with the electrical contractor. This may avoid potential problems and will save considerable time.

Furniture/Furnishings Plan, 3-F2

Shows the number and location of all selected furniture items, including furniture system components, all freestanding desks, chairs, tables, and cabinets. Although this drawing is again not specifically a construction drawing, it is produced at this point to assure precise coordination with all of the construction and panel location drawings. It may or may not be supplied to the general contractor, other than "For Your Information."

The drawing can be made by reproducing the panel plan on another blank title sheet before the panel numbers and codes have been added, and then adding all other furniture items and their respective code numbers. This drawing is separate from the furniture panel drawing for clarity. The numbering system used on this drawing is the same as that used on the furniture specifications.

Plant Location Plan, 3-PL

Shows the location and code numbers of plants, cross-referenced to the Plant Specifications. The Furniture Plan is reproduced on another blank title sheet before furniture numbers are added. The locations of the plants are indicated by a circle symbol and the plants are assigned code numbers. Again, this drawing is another of those not specifically a construction drawing. It is produced at this point to assure precise coordination with all of the construction and panel location drawings. This plan is rarely given to the general contractor.

Engineering Requirements

It is usually the function of the landlord to provide the services of the electrical, mechanical, and if necessary, the structural engineers. Who will pay for their services will largely depend upon whether the engineering work being performed is for building standard or above building standard work. This must be verified early in order to determine fee costs and to avoid last minute surprises relating to payment responsibility, as well as the more obvious problems of available power capacities, clean lines, etc. Regardless of who will pay for the services of these engineers, the company's architect should coordinate their work.

The design of most facilities requires, at minimum, two engineers: an electrical engineer and a mechanical engineer (who also normally provides plumbing engineering). It is customary to use the same engineers who acted as consultants for the construction of the building. These engineers already have an in-depth familiarity with the building and the systems they were involved in designing, and usually other building systems which have to be coordinated with their specific work. They will most likely save you time and money and be of special assistance to the company's architect because of their special knowledge. Their drawings and specifications will supplement the architectural information and be included as a part of the construction document bid set.

Request of the mechanical engineer that each private office or conference room have a separate air conditioning zone, if possible. This does not suggest that additional major air conditioning equipment need be installed. A simple variable air volume (VAV) diffuser with its own thermostat will likely do the trick. Also, each larger size conference room should be provided with a quiet exhaust fan. Exhaust fans are rated in C.F.M.'s (cubic feet per minute) and in sones. Usually the higher the C.F.M.'s, the higher the sones. Sones rate the noise level of the fan. Try for a low sone rating of below 3.0. Axial (propeller type) fans are usually noisier than centrifugal fans. Ask that each corner of building, if there are windows in the corner, have a separate zone to compensate for the differences in solar and shadow conditions. Submit final plans and specifications to the coordinator for review.

The team doing the construction drawings must provide the engineers with complete information regarding partition locations, anticipated occupant loads and use of each space. They must also supply information about electrical power requirements and loads for special equipment, a reflected ceiling plan, and all electrical, communications and data outlet locations. They should also provide both the mechanical and electrical engineers with a reflected ceiling plan reproduced at half-tone, on a title sheet so that they may coordinate their work with the ceiling grid plan. If the drawings are being done on a CAD system, the architect may be able to share disks of the ceiling plan with the engineers. The engineers will then issue coordinated duct work, mixing box and diffuser and sprinkler head locations, as well as light fixture plans with their circuiting. A half-tone electrical and telephone outlet plan should be provided to the electrical engineer, who will add and coordinate circuiting information.

Engineering drawings should be done according to the standard format in order to maintain consistency, coordination with architectural improvements, and ease of future reference.

Procurement Documents

We have assembled a complete set of the documents you will need to order or specify carpet, furniture, accessories, and plants and containers in their respective chapters. Complete instructions are included for assembling your bid and/or order packets. Carpet is more frequently ordered and installed by the general contractor. The project designer or architect may also have a good set of procurement documents. Be sure to review the architect's documents to make certain they cover all of the criteria that should be covered. Before issuing procurement documents, it is prudent to review the particular chapters in this book which deal with the items you intend to purchase. If the Architect's documents appear to be deficient in some manner, discuss your concerns with him or her. Remember it's the company's money and facility. It is better to err on the side of caution than to buy something improperly or inadequately specified.

The following is a brief description of the process and of the forms to be filled out. (Please note that if the landlord is buying the carpet in whole or in part under the lease agreement, or if the company is specifying furniture through a national buying contract, it will not be necessary to go through the complete bidding process. However, you should complete the general conditions and specifications forms.

Instructions to Bidders

An exact explanation of how bids are to be prepared, including due date, scope of work, completion time, general requirements, and preparation instructions.

Proposal Form

Stipulates how the contractor is to figure his pricing (e.g., net plus a percentage or a percentage off list). It also asks for fixed percentages or dollars for freight, delivery, and installation costs.

General Conditions

Explains basic conditions of the contract relative to the company and the contractor. The person doing the specifying must review this section to be certain all information is pertinent to the project. Additions or deletions may be made to this document with the aid of experienced or professional assistance.

Special Conditions

Contains specific information about the project, such as site location, installation dates, and drawing numbers — information to be added or filled in by the company or its agent.

Furniture and Fixture Specifications Section

Identifies items to be purchased, and can be sub-divided into general and individual sections. If the items being specified have common finishes, fabrics, construction, and/or materials, a general section is used to define all standard information on the items that follow in the individual sections.

The individual specification sheets provide information on each item, such as manufacturer's name, product number, quantity, location, and color, etc. It may also list which products will be accepted as "an equal" alternate bid or proposal. As an example, a standard format for item numbers might appear as follows in Figure 7.3.

EP .0 Series	— Electrified Panels
P .0 Series	— Panels
F 1.0 Series	— Desks
F 2.0 Series	— Credenzas
F 3.0 Series	— Tables
F 4.0 Series	— Secretarial Seating
F 5.0 Series	— General Swivel Seating
F 6.0 Series	— Guest Seating
F 7.0 Series	— Lounge Seating
F 8.0 Series	— Bookcases
F 9.0 Series	— Files
F10.0 Series	— Storage/Shelving
F11.0 Series	— Miscellaneous

Figure 7.3

In addition, consider keeping an accurate, up-to-the-minute log of all item numbers used, so as to avoid duplication of these numbers. This log should be maintained until the project is completed.

When all the forms discussed above have been filled out, bind them together with a table of contents and cover. This booklet, along with a numbered set of completed furniture plans for the new facility then becomes the official set of specification documents. The set is then forwarded to bidders.

Note that furniture items, especially systems furniture and executive furniture, require particularly long lead times, in some cases as much as 10 to 18 weeks. Therefore, as soon as the estimated number of workstations is determined, but before your final plans are complete, it is a good idea to contact the manufacturers who are supplying the major part of the furniture to reserve production time. The various dealers' and manufacturers' representatives will assist on the preparation of these orders. A letter of intent, and very possibly a deposit, may be required for this pre-production order, depending on the manufacturer and the size of the order. Once the final plans are complete, amend the order, if necessary, to indicate the exact number of stations needed.

After orders have been placed, it is important to work closely with the selected contractors and/or dealers to be sure items are delivered on time. Request copies of all confirmations received from individual manufacturers, and ask to be told of any possible delays caused by such things as strikes, and fabric or materials shortages. Watch out for discontinued items. These will occur. New selections will have to be made. Do so quickly in order to maintain the schedule. Be sure to document the re-selection with a Notice of Clarification distributed to all personnel affected.

Purchase Orders
When you are going to purchase accessories, single items, or small quantities of an item, it is not necessary to prepare bidding documents. Simply contact several suppliers to compare prices. Select one, submit a purchase order, and follow normal buying procedures.

Be sure to always:
- Stipulate the date to be delivered;
- Indicate the address to which the items are to be delivered;
- Include all manufacturer numbers, finish numbers, colors, and the like;
- Complete the pricing information, including the unit price, the total cost of freight, and all applicable taxes.

As the project nears completion, contact each vendor of materials or services. Review all commitments logged into the budget to be sure you are aware of all changes which have been made on the project. Also, obtain firm pricing on anything which has been contracted for on a cost-plus or estimate basis.

Updating the Budget

Now that all of the construction and procurement documents have been completed, do another update of the Final Budget and Schedules. Send a copy of the budget and schedules to the coordinator or officer in charge of the facility regularly, and as any major commitments are made.

Document Drawings

As the work progresses, there invariably will be changes to the work. These changes should be reflected on the drawings as revisions. Any *as-built* changes made by contractors or suppliers should be reflected on the final *hard line* drawings, either as revisions or as as-built conditions. When the project is totally complete, ask the architect to provide you with a 3 to 1 black-and-white reduction set of the complete set of fully revised drawings. (Blue-line drawings will fade.) The 3 to 1 ratio assumes 30" x 42" drawings which will produce a 10" x 14" document set. Some companies seem to prefer a so-called half-size set because they feel that this size is easier to read. However, they are harder to bind and store. You might also request a full size (30" x 42") mylar set to be used as a final record set and for use in future modifications. The objective is to get as close to an 11" x 17" size as possible. These drawings can be punched with 11/32" or 13/32" holes and bound in an 11" x 17" post binder using 5/16" or 3/8" posts.

Bind the document set in a post binder, label it with the project name, location and completion date. Bind the specifications in a three-ring binder and carefully label the binder with project name, location and date of the project's completion. If anyone needs a drawing for future work or reference, you can copy a sheet from your document set. Many in-house copy machines will handle 11" x 17" copies. Keep in mind that after a very short time the document set may be your only source of reliable information. Original hard line drawings will probably no longer be available from the original architect or designer, and your original blue-line set will have been borrowed out of existence, torn into shreds, left in the light and faded beyond recognition. Locate a convenient place to store the binders where they can be accessed only by personnel having proper authority.

Do not loan your Document Drawings. Do not write on them. Do not let others write on them. Do not lose them.

You may wish to make two or three document sets: Corporate, Facilities Management, yourself.

Chapter 8

CONSTRUCTION ADMINISTRATION

Introduction

Since lease negotiations were completed, the project has been operating either on a self-sustained basis, i.e., utilizing company staff, or in conjunction with outside architects, designers and engineers. All of these people are essentially on the company's side in terms of economic interests. From this point forward, there will be a new group of involved parties whose economic interests are not necessarily on the company's side.

The Construction Team

There has been a great deal of material written about developing a good working construction team. It is very true that having a good relationship with your contractors is most beneficial. This can be especially valuable during the preparation of the working drawings, assuming the contractor has been selected by that time. However, do not be misled, the contractor's interest in the project is to build as closely to the construction documents and schedules as possible but not to lose money. The contractor is in a legally adversarial position to the company, yet the company must depend on the contractor to construct their premises in accordance with the construction documents, within the schedule and within the budget. How successfully this is done will depend on a number of factors. First, the quality of the contractor. Second, the quality of the construction documents. Third, the reality of the schedules. Fourth, the reality of the budgets. And fifth, the realistic expectations and relations between the company and it's architectural/engineering team.

When asked what most clients want most from a contractor, invariably the response is quality of work.

The Contractor's Organization

To properly deal with a contractor it is important to understand how most contractors are organized when the work is in process. At the top of the hierarchy is the owner or president. This person often has a limited role during the actual construction. The next person is the contractor's project manager, who is often an engineer and usually responsible for more than one project at a time. Under the project manager there is the job superintendent who should be on the job full time. The "super" may be assisted by assistant supers and other staff personnel, depending on the size of the job. The super runs the job, usually using the project manager as liaison with the home office of the contractor. The principal line of communication for the company and its architects and engineers is through the super, never with the subcontractors or the job foremen. If matters are not working out with the super, your next line of communication is with the contractor's project manager, and so forth.

The Architect's Organization

Once the construction document preparation phase of the architect's work is completed, the architect will often introduce a new individual or individuals into the project in the form of the job or field architect. The field architect may also be the original project architect. This person may be responsible for more than one project at a time. He will usually visit the job daily or at least several times a week, checking on work progress, quality of work, accuracy of placement of components such as walls, shafts, electrical floor outlets, etc. It is also his job to interpret drawings, clarify ambiguities and approve any change orders. Once a month this person will determine the percentage of work completed to date and the value of material stored on the job site, and will either prepare or assist in preparing the monthly payment draw. The ability of this person to deal with the job super and contractor's project manager is critical. This architect can either find so much fault with a job so as to be constantly creating a furor, or he can solve problems.

The Company Project Manager/Coordinator

The company's coordinator or project manager is critical at this point. The company project manager/coordinator should have been closely involved with all aspects of the project as it was being developed, from the earliest considerations of new space through the preparation of construction documents. By virtue of continuous involvement, the coordinator will know the reasons certain elements were designed the way they were and how much flexibility there may be in a requested change. The company project manager/coordinator should be intimately familiar with all of the construction documents, again having been involved with their preparation. The company project manager knows whether there is room within the budget for a field improvement or substitution. The company project manager should be a frequent visitor to the site and in regular communication with both the contractor's super and project manager as well as the architect's field architect. It is the job of the company project manager to be an active coordinator and not a casual observer.

Project Protocol

Once the company has entered into an agreement with a contractor and work is about to commence, it is prudent to prepare and follow a project protocol. Once created, the project protocol should be distributed to any party having any contact with your project.

A sample project protocol chart is illustrated in Figure 8.1. It is purposely a bit more complex than what you may need. It should be modified as needed to serve your particular project. It is not the intention of a Protocol to curb discussion or to limit information from flowing between the various members of the project group. It is intended to provide a specific process for construction or contract-related communications.

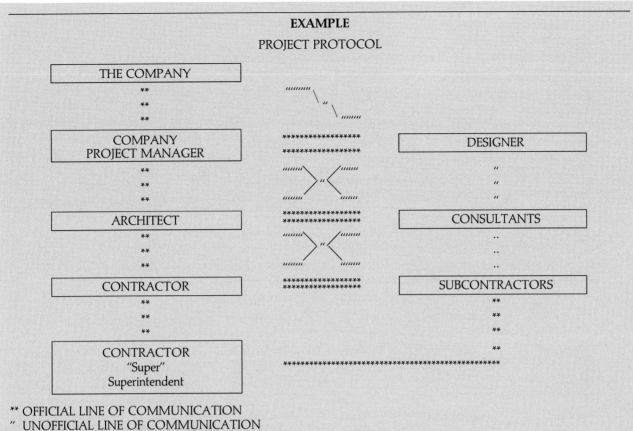

EXAMPLE

PROJECT PROTOCOL

```
┌─────────────────────────┐
│      THE COMPANY        │
└─────────────────────────┘
           **
           **
           **
┌─────────────────────────┐                    ┌─────────────────────────┐
│        COMPANY          │  ******************  │        DESIGNER         │
│    PROJECT MANAGER      │  ******************  │                         │
└─────────────────────────┘                    └─────────────────────────┘
           **
           **
           **
┌─────────────────────────┐                    ┌─────────────────────────┐
│       ARCHITECT         │  ******************  │      CONSULTANTS        │
│                         │  ******************  │                         │
└─────────────────────────┘                    └─────────────────────────┘
           **
           **
           **
┌─────────────────────────┐                    ┌─────────────────────────┐
│       CONTRACTOR        │  ******************  │     SUBCONTRACTORS      │
└─────────────────────────┘                    └─────────────────────────┘
           **                                              **
           **                                              **
           **                                              **
┌─────────────────────────┐                               **
│       CONTRACTOR        │  ***************************************************
│         "Super"         │
│     Superintendent      │
└─────────────────────────┘
```

** OFFICIAL LINE OF COMMUNICATION
" UNOFFICIAL LINE OF COMMUNICATION
.. UNOFFICIAL LINE OF COMMUNICATION – TRY TO AVOID

Notes:

1. If at all possible, use *official* lines of communication. Example: A subcontractor wants to confirm a specification condition with the company requiring a 24-hour notice of a power outage. The subcontractor should call the contractor, who should call the architect who should call the project manager, who will notify the company. If the communication is written, it should be routed to each party using an initialled italics box for confirmation.

2. If time or circumstances do not allow normal communication, then the unofficial lines should be used, followed by copies or contacts through official lines.
Example: The company discovers pressure grout leaking into an electrical vault. The company calls the general contractor and then the project manager who, in turn, calls the architect.

3. The object is not to stop or limit communication, but to ensure that everyone who needs information gets it.

4. Even in non-emergency situations it may be desirable to use unofficial lines of communication. When this is done, a mutual understanding should be made as to who will follow-up through the official lines.

5. Lines other than those shown should not be used even under unusual circumstances. This type of communication could cause serious contract problems.

6. Submittals or change orders which require review and signature must move up through official lines, but once signed should be transmitted directly to the contractor, unless the item is modified or rejected.

7. Information or instructions flowing down the chart for signature should always move through official lines in some form. Revisions or exceptions should also move back up through official lines.

8. Submittals through official lines which are different from the specifications or drawings should be so noted. Prior notice of such a condition through unofficial lines may avoid needless rejections and/or paperwork.

Figure 8.1

Once the construction documents have been completed, they are sent to the company's project manager for use in contractor selection, if appropriate. At the same time, a copy may have to be sent to the landlord (if he requires a set of the drawings for his approval before any work is started).

Schedule

If the company's landlord is providing the contracting services, give him and his contractor a copy of the project schedule. While a schedule was probably developed during the lease negotiation process for the purpose of establishing the *rent commencement date*, it is unlikely that you will be able to confirm your project schedule until the construction drawings have been completed or nearly completed. If the landlord's contractors are unable to keep to the new schedule the landlord should extend the rent commencement date and should be responsible for handling all problems if the company must vacate its present premises by a certain date and that date occurs before it can move into the new space.

If the company is bidding the construction work itself, give each bidder a copy of the most current schedule. In particular be sure that your most current schedule will meet the rent commencement date. If a bidder's estimate is predicated on a longer time period than that allocated in the schedule or requires overtime work, you may have to disqualify that bidder, unless your rent commencement date can be extended without charge.

When Construction Bids Are Not Necessary

In most cases the Tenant Work Letter in your lease agreement will have been negotiated to include much, if not all, of the construction necessary to complete the company's facility. Usually, the contractor retained by the landlord will construct the normal tenant improvements such as walls and doors, ceilings, lighting, electrical outlets, and floor coverings.

In addition, that same contractor will most likely construct the above standard items, which are those things not usually provided by the building owner. These items include cabinetry, glass partitioning, dumbwaiter, inter-floor stair, special wall covering, shelving, etc. In most cases, the charges for these items will not warrant the time and expense of going through a complete and separate bidding process, and the coordination between two different contractors would be prohibitively difficult.

However, before arranging for the landlord's contractor to do work for which the company will be paying, verify that other tenants in the building have found his work satisfactory, that his fees are competitive in the marketplace, and that his proposed charges are comparable to what he has charged other building tenants. Also find out how he will invoice for his work. Evaluate the best method of paying for such construction: in monthly progress payments or in one payment at the completion of the work. You may find that your landlord will agree to amortize the cost of such construction over the term of your lease. Check with the officer in charge of the facility or the Coordinator to determine if this is a desirable alternative. The likelihood is that all of the above items were worked out during the Tenant Work Letter and lease negotiations. If not, read the sections in this book concerning these matters and attempt to establish similar arrangements with the landlord and his contractor.

When Construction Work Should or Must Be Bid

As an alternate to bidding one general contractor against another, you might follow the selection and approval process described below for the selection of one general contractor and then having him competitively bid the work of the subcontractors. To do this it will be necessary to arrive at an agreement with the Contractor on the costs and percentages he will use for such things as general conditions, overhead, profit, unit costs, for work not performed and for work added, etc. This is a particularly helpful practice when you are operating on a tight schedule. In addition, the company may be able to have the contractor on board during the preparation of the construction documents to assist in doing value engineering and to provide the current empirical knowledge of more efficient or less costly ways of constructing certain portions of the project.

Select a general contractor through either a bidding or negotiation process when:

- The company is renovating its existing space, but the landlord is not contributing to that remodeling;
- The landlord does not have a particular general contractor who works in the building;
- The company is leasing new space without a tenant improvement allowance and it will pay for all of the improvements;
- The landlord is making a cash contribution which will used to construct the company's space.

To begin the bidding process, schedule individual interviews with several contractors and thoroughly review their qualifications.

One of the best times to do this is after completing the *design development drawings*. The design development drawings, at this point, represent a reasonably inclusive description of the project, without all of the details finalized. Using the design development drawings, ask each of your short-listed contractors to give you a reasonably accurate estimate of the job cost and an indication of the schedule required to build it out after the final *working drawings* or final construction documents are completed. Advise the contractors that their estimates will not be taken as bids, but only as indications of the cost of the work as described in the design development drawings. However, advise them that should they be selected for the project, their estimates will be used in arriving at the final contract amount. Also ask each contractor to make available to you his line item estimates should he be selected. By not asking for the line item estimates *prior* to making your selection, the contractors will not be as concerned that you may be peddling their bids. In other words, telling contractor A how contractor B's line-item estimates compare with his. Yet, you will have access to the selected contractor's line-item estimates during the preparation of the working drawings. This information permits the architect and the engineers to design out costs which may be pushing any costs over budget.

In the interview:

- Review the contractor's familiarity and experience with projects of similar type and size to yours.
- Ask for the name, recent past experience, and references for the specific superintendent the contractor would use on the company's project. This person can make or break both the budget and schedule.
- Discuss the contractor's billing procedures. For example, will he charge a fixed fee, a percentage of the total construction costs, or make some other arrangement? What is his procedure for charging for changes to the work after construction has begun?
- Ask for the names of and speak to approximately three previous clients for whom the contractor has done work similar to yours.

Particular points to check with these references include:

- What is the contractor's reputation for completing a job with minimum change orders? What kinds of changes were involved?
- How did the total dollar amount of change orders compare to the original bid?
- What was the contractor's record for completing projects on schedule?
- How did the actual completion date compare to the originally scheduled one? In the case of delays, were there any mitigating factors over which the contractor did not have any control?
- What was the quality of work?
- How cooperative was he; how responsive to questions or concerns; and to non-time-consuming or non-economic changes?
- Ask the prospective contractor to submit a letter from his bonding agent describing his current bonding rate and the amount of *open* bond capacity currently available. This information will give a good feeling for the current financial strength of this contractor.

If You Are Doing Conventional Competitive Bidding

Once the contractor interviews are completed and their references checked, select no fewer than two and no more than four contractors and request their bids, to be based on the completed construction drawings of your project. The project architect should provide a bid form for your particular project.

Depending upon the size of your project, allow a minimum of two to four weeks for the contractors to prepare their bids. Do not rush this process. The contractor needs time to contact his sub-trades and to put together his estimates. Rushing him will cause him to guess as to costs, and he will doubtless guess on the high side.

At the beginning of the bidding process and again approximately one week before the bids are received, schedule a meeting with all of the prospective bidders, to review and clarify any questions they might have. There is no value in having contractors bidding on work which may be unclear to them. There is also little value in asking for bids from contractors with whom you or your project team members are uncomfortable.

When bids are received, review and compare them to be sure that each bid includes all phases and aspects of the project, and that there are no exceptions or allowances which are not firmly committed. When you are satisfied that the bids are comparable in scope and form, forward them or a summary of the results to the officer in charge of the facility, if necessary, with your comments and recommendations, for final review, selection, and drafting of the formal owner/contractor construction agreement.

Your contractor should procure all necessary building permits, as well as permits necessary from the fire department, health department, and any agencies regulating building standards for the disabled. Usually no special permits are needed beyond the basic building permits. However, you will often have to file a cash bond with the city to cover future costs of inspections to be made by the building inspectors. Once the work is completed and the building inspector has issued the permanent certificate of occupancy, any remaining balance in the cash bond will be refunded. You will pay for the permits either as a separate charge or as a part of the contractor's general conditions. Check to find out where this cost is being carried; if the contractor has applied any form of mark-up on it. You do not want a surprise in the form of an extra.

Construction Administration

The company project manager (or other person in charge of administering construction for the company's project) must maintain close contact with the contractor and his job superintendent until every aspect of the project is completed. The responsibilities involved are discussed below. The various forms cited are illustrated as completed examples in Figures 8.2-8.8. Full-size blank forms that may be copied can be found in the Master Forms Section.

Job Meetings

Set up and conduct regularly scheduled meetings (weekly is best) to coordinate with the work of the contractor and any subcontractors or consultants and to maintain a feel for progress, delays, problems, etc.

Record your own meeting minutes of these meetings. Do not rely on others to maintain your minutes. The minutes they prepare may not reflect your concerns or what you understood to have been said or agreed upon. If you feel that there were matters of concern agreed upon or discussed during the meeting, distribute copies of your minutes to the contractor so that he may take any required actions. Number each item brought up at the meetings in a straight numeric sequence. Once placed in the minutes, it remains there referenced by meeting dates with a description of what action or disposition was made of it on each meeting date. The item reappears in each meeting minutes with comments as to what action was to have been taken and by whom until the item is finally resolved, after which it can be removed from the minutes.

Job Visits

The company's project manager or other representative should make periodic visits to the project site to become thoroughly familiar with the progress and quality of work and to determine conformance with the construction documents. If any discrepancies are found, advise the architect immediately so that he or she may advise the contractor in writing. Keep an on-site visit log (in the form of a memo), with a copy going to the architect and the contractor. For most projects, your architect will provide professional assistance in determining that the work complies with the construction documents.

Site visits are also necessary to accurately determine the payments due the contractor, which will be based upon the architect's observations of the amount of work completed and of materials stored at the site.

It is extremely important to remember that the company as a client of the contractor does not have the authority to advise or issue directions concerning construction methods, techniques, procedures, schedules, safety prescriptions or programs in connection with the Work; or to personally conduct any tests. If there are matters you feel should be questioned or brought to the contractor's attention, do so, but do so using the proper protocol. First, contact the architect, designer, etc.; if the company has none, contact the contractor's project manager. If you feel your concerns merit extraordinary urgency, contact the job superintendent. Never give instructions to the workers on the job. They work for the contractor and the contractor works for you. Workers like nothing better than to confuse or delay (*dog*) the job by telling their own boss that "...so-and-so said."

Revisions to Drawings or Specifications

The only document that can legally change the contract documents once the bid has been accepted and the owner/contractor agreement has been executed is a change order.

Any requests for changes or clarifications in the contract, whether made by the company or by the contractor, must be done using this form, which must be signed by both the company and the contractor. In the event that an architect is involved with the contract documents, he must be the first to approve and sign the change order, and will probably actually prepare the change orders since he is legally responsible for administration of the documents. A construction administration log should be kept to record all changes. Examples of a filled-in change order and a filled-in construction administration log are illustrated in Figures 8.2 and 8.3.

When a change or clarification is required, it must be documented by one of the following forms:

Notice of Clarification

The *notice of clarification* simply clarifies an item when the intent has already been set forth. It does not involve a change of either contract amount or completion time. A completed example of a notice of clarification is shown in Figure 8.4, and a blank form is found in the Master Forms Section.

Estimate Request

If an actual change or addition to the contract documents is desired, a price must be solicited from the contractor through an estimate request. The contractor will then determine the change in the contract amount caused by this change and the amount of extra time needed (if any) to complete the changes, and return the information to the company project manager, through the architect. If you feel that the information is in order, you will then issue a change order to authorize the contractor to proceed. Figure 8.5 illustrates a filled-in example of an estimate request. A blank form is found in the Master Forms Section.

Field Order

If a requested change requires immediate action, and the probable cost and time effects are nominal, a field order authorizing the contractor to do the work before he states the difference in amount or time needed, may be issued by the architect and countersigned by the project manager. After these costs and time requirements have been determined, a change order will confirm the field order. Issuing field orders which affect the contract amount or schedule is a dangerous practice and is not recommended. Figure 8.6 illustrates a filled-in example of a field order; a blank form is found in the Master Forms Section.

Shop Drawings

The company project manager, architect or designer must review all shop drawings for compliance with contract documents and the design intent. shop drawings are detailed drawings prepared by a contractor, his sub-contractor or suppliers, which indicate the methods of fabrication of specific components of the construction. These shop drawings should be logged in a shop drawing log maintained in accordance with the Uniform Construction Index. An example of a filled-in shop drawing log is illustrated in Figure 8.7. A blank form is found in the Master Forms Section.

Change Order

No. 2

Project: Branch
Location: Fairview, IL

To: Geggatt Co.
Date: 12/20/94

The following change to the contract is effective only upon signature by all applicable parties.

Description/References	Change in Contract	
	Cost	Time
Substitute 2'x2' fluorescent fixtures for all 2x4 fluorescent fixtures (1124 @ $6.52)	7328	Ø
Add 20 2'x2' fluorescent fixtures & install $74.30/Each	1486	+ 1 Day
Modify stair 'A' banister per sketch	922	+ 2 Days
	9736	3 Days

Current contract amount $ 611,500   Completion date: 9/30/94

Add (Deduct) $ 9,736   Add (Deduct) time: 3 Days

Revised contract amount $ 621,236   New completion date: 10/3/94

Recommended by: D. Johns   Date: 2/20/94

Architect's approval: Pat Boyle   Date: 2/21/94

Accepted by contractor: Allen Geggatt   Date: 2/24/94

Accepted by owner: _____   Date: _____

Accepted by tenant: _____   Date: _____

Figure 8.2

211

Construction Administration Log | Change Order Log

CONSTRUCTION ADMINISTRATION LOG / CHANGE ORDER LOG

Facility: **Branch** Location: **Fairview, IL.** Page ___ of ___

REFERENCE NO.	DESCRIPTION	REQUESTED BY	REFERENCE DRAWING	CONTRACTOR DATE SENT	CONTRACTOR DATE RETURNED	AMOUNT	TIME EXTENSION	REFERRED TO	REFERRED DATE SENT	REFERRED DATE RETURNED	APPROVED	REJECTED	CONTR. DATE SENT	CONTR. DATE RETURNED	OWNER DATE SENT	OWNER DATE RETURNED	REMARKS	DATE	CONTR.	FIELD	OWNER	FILE	CHANGE ORDER NO.
	Structural Drawings	Cochran	S/8	12/3	–																		
F.O. 1	Modify Stair 'A'	D. Johns	A, C	12/5	12/8	$922	3	Arch.	12/5	12/8	✓												
E.R. 5	Substitute 2x2 for 2x4 lights	D. Johns	E	1/15	2/7	$320	0	Elect.	2/24		✓		2/23	2/24									No. 2
E.R. 5	Add 20 2x2	D. Johns	E	1/15	2/27	$486	1	Elect.	2/24		✓		2/23	2/24									No. 2
NC 1	Relocate Stats	D. Johns		12/10		–	–	Mech.	2/10	✓	✓		2/23	2/24									No. 2

Figure 8.3

212

Notice of Clarification

No. 1

Project: Branch

Location: Fairview, IL

To: Geggatt Co.

Date: 12/10/94

The following items are clarifications to the drawings and/
or specifications. They should NOT INVOLVE A CHANGE
in contract price or completion time and shall be
incorporated immediately.

All thermostats shall be mounted at 42" from F.F.
rather than at 72".

By: David Johns

Figure 8.4

Estimate Request

No. __5__

Project: __Branch__   Location: __Fairview, IL__

To: __Geggatt Co.__   Date: __1/15/94__

The following change in work is being considered. Please prepare a quotation showing the change in cost for each item, the distribution of costs, and the change in completion date, if any. Promptly return a copy to the architect's office. Work shall NOT PROCEED prior to issuance of written instruction unless specifically authorized.

1.) Substitute 2'x2' lightolier CD2 parabolic louver fixtures for

2'x4' lightolier CD4 parabolic louver fixtures

2.) Add 20 additional fixtures installed.

Reason for change: __Ceiling design changed__

Submit estimate no later than 2 weeks or by: __2/7/94__

By: __David Johns__

Figure 8.5

214

Field Order

No. 1

Project: Branch
Location: Fairview, IL

To: Geggatt Co.
Date: 12/4/94

The following instructions are intended to clarify and expedite the work. If a change in the contract amount or time is entailed, a Change Order must be executed.

[✓] Proceed immediately   [   ] Do not proceed until:

Proceed with modifications to stair 'A' banister A.S.A.P. per attached sketch. Provide pricing by 12/15/94.

Send to Hall Assoc. Architects. Attn: Pat Boyle
Change order will follow pricing.

Reason for order: Owner's request for design change

By: David Johns

Figure 8.6

Shop Drawing/Sample Log

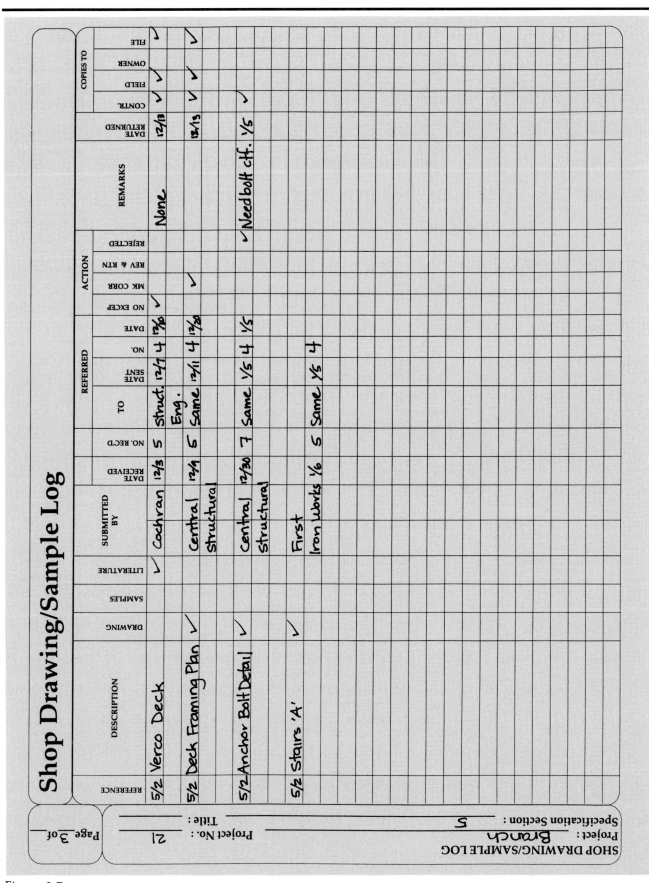

REFERENCE	DESCRIPTION	DRAWING	SAMPLES	LITERATURE	SUBMITTED BY	DATE RECEIVED	NO. REC'D	REFERRED TO	DATE SENT	NO.	DATE	NO EXCEP	MK CORR	REV & RTN	REJECTED	REMARKS	DATE RETURNED	CONTR.	FIELD	OWNER	FILE
5/2	Verco Deck	✓		✓	Cochran	12/3	5	Struct. Eng.	12/7	4	12/10	✓				None	12/13	✓	✓		✓
5/2	Deck Framing Plan	✓			Central Structural	12/9	6	Same	12/11	4	12/20		✓				12/13	✓			✓
5/2	Anchor Bolt Detail	✓			Central Structural	12/30	7	Same	1/5	4	1/5					✓Needbolt ctf. 1/5		✓			
5/2	Stairs 'A'	✓			First Iron Works	1/6	5	Same	1/5	4											

SHOP DRAWING/SAMPLE LOG

Project : Branch

Project No. : 21

Title :

Specification Section : 5

Page 3 of

Figure 8.7

216

Completing the Construction

When the construction is nearing completion, review the project in the presence of a representative of the contractor. Carefully prepare a *Punch List* (a list of those items yet to be completed and/or corrected). All work noted on this list must be finished before final payment can be made to the contractor. In order to prepare the list, the team preparing the punch lists will need a set of the construction drawings and a punch list form for each area you are reviewing. Figure 8.8 shows a filled-in example of a punch list; a blank form is found in the Master Forms Section.

When preparing the punch list, the company project manager, the architect and a representative of the contractor work as a team reviewing every space within the project. After filling out the information at the top of the punch list form, and using the construction drawings as a guide, start at one end of a floor and work your way completely around that floor. It is usually a good idea to do this in a clockwise manner to be sure you do not overlook anything.

As a point of reference, list each room or area number (check the construction plans for those numbers). Then list all items within the area that must be corrected and/or completed. Try to group these items on your punch lists by the subcontractor responsible for them. For example, list all electrical problems together, all carpet problems together, and so on.

A good way to observe a room in checking for problems is to run your eyes along all corners (between wall and ceiling, wall and floor, and where two walls intersect).

Remember that the punch list(s) are not limited to work done solely by the general contractor, but should include any subcontractors' work as well. Therefore, ask the contractor to assign the appropriate subcontractors' names to the follow-up column of the form. Prepare a separate punch list for furniture and equipment-related problems for each vendor not under contract to the contractor.

Once the punch list(s) are finished, issue them to the contractor, with copies to the landlord, if he is responsible for the construction. The contractor will distribute the punch lists to any subcontractors involved. Include a deadline by which all items must be corrected and/or finished, usually one week to one month, depending on the complexity and size of the project. After that period of time has passed, do a second walk-through to review the punch lists and check off those items that have been corrected. If necessary, the whole process will have to be repeated until all items are deleted from your punch list(s) and construction is completed to the company's satisfaction.

Only after all work has been corrected and completed to the company's satisfaction should it release the balance of the "holdbacks."

Holdbacks

It is a customary practice for the *Owner*, in a contractor or supplier relationship, to hold back a portion of all the funds approved for payment to a contractor. The customary and usual amount is 10% of all approved draw amounts. These funds are usually retained until completion of the project or installation of all furnishings and furniture.

There are, however, some exceptions to these generalizations. When a project is effectively completed and there remains only punch list work, whether construction related or furnishings related, it is normal practice for the architect, in conjunction with the company project manager, to evaluate the costs of corrective work and hold back amounts equal to the estimated costs of the corrective work. All other holdback funds are released to the respective

Punch List

No. _____

Project: **Branch**   Location: **Fairview, IL**

Floor: **4th**   Unit: _____   Room: **408**

To: **Geggatt Co.**   Date: **1/10/95**

Prepared by: **David Johns**   Phone: **555-8989**

Items to be resolved. The following items of work are either incomplete or improperly completed.	Contractor follow-up by:	$ Amount withheld	Date to be completed by:	✔ Completion
Double switch lighting per drawings	Felton Elec.	$ 150		
Supply & install cover plates on outlets	Same	25		
Replace all "cool white" lamps with "warm white" per specs	Same	100		
Repaint north wall – poor quality job –	Morris Painting	100		
Touch up closet interior per "touch up" allowance		25		
Repaint south wall with correct color per finish schedule 2 coats, or as needed		250		

Figure 8.8

contractors and suppliers. It is not the function of the holdback fund to be punitive, but to reasonably cover incorrect, inadequate or incomplete work or materials.

Schedules and Completion

Do not be surprised if the scheduled completion date appears before you feel ready for it. Experience shows that the most critical construction control period is in the middle of the schedule. During the early construction portion of the project, there will be a *getting acquainted* period. It is the time one learns which construction subcontractors or trades will exercise the best control and which will not. One discovers how good the super and the general contractor's and architect's project managers are at managing and administering. Inquire about deliveries of critical materials, especially those which have to come directly from a manufacturer, rather than out of warehouse stock. If the schedules are not being met, voice your concerns to the architect, and if you feel that a more direct effort is required, discuss your concerns directly with the general contractor's project manager. Try to be specific, but also try to be fair. You may wish to begin by expressing your concerns and inquiring as to what they, the project managers, feel the problems may be.

If the problems persist and it appears that the original schedules may not be attainable, you may need to review your options. These options may range from requesting that the contractor(s) put more personnel on the job, substitutions for delayed materials, insisting on or authorizing overtime, phasing the job to complete some portions to allow critical groups to move-in on time, to re-scheduling your move-in date. Always keep in mind that the client company is the boss. It will be most affected by a delay. Try to assist in problem solving, but also try to press for full compliance with the schedules.

Chapter 9

INSTALLATION AND MOVE-IN

Introduction Final installation is a step-by-step process which begins after the wall finishes are complete, the carpet is down, the doors hung, the ceilings and light fixtures installed and the lighting, air-conditioning and heating systems are all operational. The company's facilities manager or coordinator will be responsible for coordination between the general contractor and those subcontractors and vendors over which the general contractor has no control (such as furniture dealers). Although the exact timing of each step will depend on the size of the project, the work will usually be done in the following order:

Carpet

To ensure an on-time move-in, base the timing on the amount of time the carpet installers will need to complete each area and the time available before furniture must be installed. In normal situations a crew of carpet installers can average approximately 2,000 square yards per week. This equals 18,000 square feet, or about an average size floor.

Woven construction carpet should be directly glued to the floor because it is used on large open areas under free-standing furnishings and chairs with casters. The manufacturers recommend a padless installation in order to minimize wear and stretching caused by the movement of the seating. With proper installation, better quality carpets will be guaranteed, for wear only, by the manufacturer for up to 10 years.

It will take several days for the carpet to adhere properly after gluing. Therefore, if the furniture is being installed immediately after the carpeting is laid, the contractor or the mover must cover the carpeting in the main traffic aisles with plywood or hardboard so that the movement of heavy dollies will not break the glue bond.

Workstations

If you are using systems furniture workstations, note that they are composed of panels of different sizes to which components are to be attached. Raceways within the panels are available to carry the electrical power, telephone and data systems cabling to the various stations within a section.

Properly installing the workstation system is a complex process and should be done by a manufacturer's authorized installer. Usually this work is handled by your local dealer and should be done under the supervision of your coordinator. Installation is basically a five-step process:

It is critical that the placement of panels be carefully checked against your panel plan. Some panels may look the same, but differ by as little as 2". A seemingly insignificant difference will become a major problem when the time comes to install the components. Follow the panel plan which was prepared either by the architect, designer or furniture dealer and should be a part of the construction documents.

Some systems furniture panels are pre-wired; most are wired with factory-produced cables or harnesses after the panels have been erected. A connection is then made to the building power source. This work should always be done by qualified electricians. These may be employed by the contractor doing the construction, by the landlord or by the company. Be careful of jurisdictional disputes.

Telephone, computer and other cabling is run through the panel raceways, then the telephone company crews will connect the telephone sets and make the connections to the telephone switch.

Since many (if not most) workstations are connected to a computer system and will have a video display terminal, they must have their computer cables installed in the raceways along with the telephone cables.

The systems furniture components are installed. If task lighting fixtures are part of the plan, the components using these fixtures are to be installed first, as additional electrical work may be needed to connect the lights. When all necessary electrical work is completed, all telephones are hooked up and computer cables are installed, the systems furniture components and work surfaces are installed according to the furniture plan.

Storage and Filing Equipment

If the filing and storage areas are centralized, as they are in many facilities, and thus not in the same area as the workstations, storage and filing installation can be done at the same time as the workstations. If, however, your furniture system and your shelving and filing system areas are close together, we suggest installation be scheduled at different times to avoid unnecessary congestion and freight elevator conflicts.

Other Furniture

Chairs, tables and free-standing cabinets are the last pieces of furniture to be put in place. Wait until just before move-in, as late as the afternoon before to install any miscellaneous items such as desk top accessories and wastebaskets. Sometimes it is best to not distribute accessories until two or three days after move-in.

Plants and Artwork

If at all possible, plants and graphics should be in place before the facility begins to operate. First impressions are long-lasting ones, and every effort should be made to make them positive. Artwork should be installed *after* move-in.

Several Points to Remember

When supervising the installation process, be sure to:

- Arrange for the use of the freight elevators in both the present and new buildings well in advance, and indicate how long they will be needed.
- Carefully check the number of items delivered to be sure it agrees with the quantities ordered. If there is a discrepancy, notify the supplier at once.

- Examine all furniture and equipment for damages. If any damages are found, tag those items, set them aside, and contact the trucking company and the supplier immediately.
- Protect walls and floors in all main traffic areas to avoid possible damage when pieces of furniture are being moved in. The mover or contractor can usually supply corrugated cardboard and hardboard for this purpose.
- Try to schedule installation so that working departments are disrupted as little as possible.

If major construction is involved, try to have it completed area by area or floor-by-floor so that the move into the new space is done in phases rather than all at one time.

If renovations are going on while employees are working, inconveniences such as noise and dust cannot be avoided. Even though builders can usually control dust and dirt to a large extent, company employees will still have to be asked to be patient. Remember, most employees take pride in their work. They will often view interruptions as causing them more work and affecting their work quality. Extend some understanding for their concerns.

Selecting a Mover

If the company is planning to move between June and September, it may have to select its mover four to five months ahead of the move date, as these are the busiest months for most moving companies. If the move is scheduled for any other time of the year, two or three months advance notice should be enough lead time. The amount of lead time may vary depending on the region of the country.

Obtain proposals from several movers and compare costs and references. It is absolutely vital that you check references carefully to analyze past performances, particularly to learn how the final moving costs compared to the original estimate. Never select a mover on the basis of cost alone. Often the moving company with whom you are dealing is merely a broker for the owner of a truck or for a driver and crew. The people doing your move may not even be employed by the moving company with whom you have negotiated your contract. Also, be careful about damage insurance. The standard ten-cents-per-pound coverage is based upon the weight of each item. A 35 pound computer is insured for the same dollar value as a 35 pound chunk of coal. If the company's move is large, consider hiring a moving consultant to help you contract and organize the move. Occasionally a mover will offer you a rock-bottom price in order to have the reputation of handling a big move for a name client. Do not let a mover practice his trade on your company. It can end up costing a great deal of money and aggravation.

Specialty movers may be desirable or necessary to move such things as computers and artwork. Select these movers with the same care as the one for the general move. Often the mover selected for the general move will have a special division for moving these items. It will be simpler and perhaps less costly if all aspects of the move can be combined under one moving contractor.

More Caution

In some sections of the country mover's charges are regulated by the state, which allows the mover to be paid for its time, material and other costs. In this case, your problem becomes much more complex. You must determine the mover's exact charges for each of his different job classifications, e.g., drivers, movers, packers, foremen and what the foreman can and cannot do.

Who can operate the elevators? What will be his job classification and rate? This assumes you are required to have an operator. If the building has not been turned over to the owner, neither the mover nor the owner may be able to furnish the elevator operator. It is possible that the contractor may have to do so.

Try to determine the charges for trucks and other equipment such as radios, racks, dollies and other moving and lifting devices, pads, blankets, runners, etc. Are the charges by the hour, the day or the job? If by the hour, when do the charges begin and end? Are there charges for mileage in addition to, or in lieu of the time-related charges?

Next, require the mover to provide his crewing schedule. How many people are to be used and where? How many drivers per truck? How many trucks? How many people in the crews are assigned to loading and unloading the elevator? How many floor movers preparing furniture and taking it to the elevators and from the elevators to the specific place to which each item is assigned? How many packers?

Ask for a justification of the crewing schedule. Is it enough to do the job within the budgeted time? At what point will overtime kick in, for which job classifications, at what rates, and what are the limits? Is the crewing schedule too large and is the job being featherbedded? Or is it too light in order to force overtime?

Are driving times to the new site adequate or too long, and is there to be a turn around time?

What are the rules regarding breaks and meal times? Will a truck be required to stop in mid-transit for a coffee break or noon meal while loading and unloading crews just sit?

What are the costs for packing materials—boxes and barrels, by size and type? Are any of them returnable, if so, at what rates?

Even if your move is of a firm contract type and not time and materials, it is beneficial to get answers to most of the above questions in order to provide the ground rules in the event of overtime or other forms of extras. As you can see, the reputation and integrity of the mover is extremely important.

The check list in Figure 9.1 will assist you in choosing the mover best qualified to handle your job. Copies are found in the Master Forms Section.

Question number two on the check list deals with union affiliation, which may or may not be important in your building. In some jurisdictions union construction workers working late into the job, such as telephone installers, HVAC balancing technicians, etc., may walk off the job if you are not using union movers and installers. In other areas this will not occur. Ask about union affiliation to verify if it is important in your area and if the workers doing construction work in the building are working for contractors with union agreements. If union affiliation is important, be sure to anticipate it and to plan for it.

When the project has been simply remodeling or an addition to existing space and the move is not a large one or out of the building, it may not be necessary to use a professional moving company. If one department is simply moving from one location to another in your present facility, some of your own staff may be able to do the physical moving after hours or on a weekend. Or, you may find that the porters in your building are experienced in handling in-building moves. While there will be charges for their services, you may find that they are as skilled as a professional moving company and probably significantly less expensive. In any case, still give your employees detailed instructions on pre-move housecleaning, labeling and packing.

Facility: _____   Location: _____

Prepared by: _____   Phone: _____ Date: _____

Moving Co.: _____

Address: _____

City: _____   State: _____ Zip: _____

Representative: _____   Phone: _____

Number and type of vans owned: _____

_____

**Length of full-time service,
major projects supervised or
worked on in the past two
years; and union affiliation of:**

     1. Management Personnel:

     2. Key Men:

     3. Rank and File:

Years Service	Projects	Union Affiliation

Capabilities (describe):

A. Move planning services--special methods and procedures: _____

_____

_____

B. Special types of equipment and containers available: _____

_____

_____

C. Warehouse and garage facilities: _____

_____

_____

_____

Figure 9.1

D. Assembly service for new furniture: _____

_____

_____

_____

E. Furniture repair and refinishing: _____

_____

Mover's own shop: What equipment and craftsmen are available? _____

_____

If not, how are damages handled? _____

_____

F. Can the mover dispose of surplus furniture? _____

_____

G. What commitments has the mover scheduled immediately before and after scheduled move date?

_____

Cost and Billing:

A. Stipulate tariff-chargeable items--unit and unit rate (e.g., labor @ $ _____ per hour): ___

B. Stipulate non-tariff items and rate (e.g., storage, demurrage, insurance, etc.): _____

C. Total cost estimate: _____

D. Payment terms: _____

Insurance coverage carried by mover (e.g., hold harmless, Workers' Compensation, personal liability, property damage, etc.)?

_____

_____

_____

_____

Figure 9.1 *(continued)*

References (name and phone no. of reference, move date and scope of job). List 3 and interview each; write all comments below:

_____

_____

_____

_____

_____

_____

_____

_____

_____

_____

_____

_____

_____

_____

_____

_____

Other comments: _____

_____

_____

_____

_____

_____

_____

_____

_____

_____

_____

Figure 9.1 *(continued)*

Preparing for the Move and Moving In

Whether the company's move is a major one to a new location or simply from one floor to another, it is important to follow some basic procedures and to do some advance planning. This holds true whether you are re-using existing furniture, buying new furniture, or combining new and existing furniture. Other managers who have recently moved, can often give helpful ideas and suggestions on how to cope with a move.

When making a major move, first select the mover as previously discussed. Since each moving company has its own particular system for expediting a move, the mover selected should furnish detailed planning and organizing instructions. Nonetheless, do consider the following points:

At the Existing Location

1. Coordinate with your present landlord to:
   Secure a building security release.
   Clear restricted moving times.
   Reserve dock space and time.
   Reserve elevator time.
2. Coordinate with:
   Phone companies (Utility and Installation).
   Electric utility.
   Other outside services such as Federal Express, UPS, etc., and the Postal Service, all of which will be affected by your move.
3. Arrange for mechanics, carpenters, plumbers, electricians, etc., to disconnect equipment as necessary.
4. Will special traffic control be needed at street loading areas?
5. Arrange for packers,
   in-house personnel and/or
   the mover's personnel.
6. Arrange for mechanics to disassemble furniture, such as desks, shelving, etc.
7. Select personnel to assist the mover in labeling operations.
8. Arrange for labeling, disconnecting, packing, distribution and connection of equipment items such as telephones, VDT's, fax, and telex machines.
9. Select personnel to assist in supervising the actual move.
10. Select personnel to assume custody of sensitive documents or contents in a vault or safe, or other valuable material.
11. Coordinate move with security personnel.
12. Arrange with current landlord:
    To deliver vacated space.
    Secure landlord's release.
    Deliver keys.
    Arrange for return of security deposits.

At the New Location

1. Coordinate with your new landlord to:
   Clear restricted moving times.
   Reserve dock space and time.
   Reserve elevator time.
2. Coordinate with:
   Phone companies (Utility and Installation).
   Electric utility.
   Other outside services such as Federal Express, UPS, etc., and the Postal Service, all of which will be affected by your move.

3. Confirm that the new telephone system is operational and reliable prior to move-in. Confirm that all PC, VDT cables and LAN cables are in place and operable.
4. Will special traffic control be needed at street unloading areas?
5. Arrange for mechanics, carpenters, plumbers, electricians, etc., to reassemble equipment as necessary.
6. Be sure that selected personnel assisting in supervision of the moving-in process are on location and have transportation to the new location.
7. Arrange for distribution and connection of equipment items such as telephones, VDT's, fax and telex machines.
8. Make available keys to locked equipment and to office and entry doors, as needed.
9. Log in valuable and sensitive material and place in new vault or secure area.
10. Coordinate the printing and distribution of new stationery, business cards forms, etc..

Timing

1. Have you coordinated your move dates with all necessary parties?
   - The moving company.
   - The current landlord.
   - The new landlord.
   - The company's personnel.

2. Have you considered the effect of down time?
3. Have you arranged for special servicing of equipment such as:
   - Typewriters
   - Fax machines
   - Copiers
   - Vending machines
   - Computers, etc.

4. Is there a need for a set of backup dates in the event of construction or furniture delivery delays?

Personnel

1. Provide instructions to employees and to the general public (change of address notices, etc.); (See Get Ready to Move-in, Figure 9.2 and in the Master Forms Section.) (See Orientation Memo and suggested map.)
2. Establish a firm moving date as soon as possible. Such preliminary moves as dismantling and re-erecting shelving, supply bins, and stock areas usually takes several days, both to dismantle and to reassemble.
3. Schedule planning and orientation meetings with company coordinators and with the moving company representatives.
4. Inform company personnel of the moving dates as soon as possible – and keep them informed.

Other

1. Once the move date is set, work with the mover to prepare a tentative schedule. He may suggest some scheduling options, such as moving some departments on one weekend and others on a following weekend.
2. Inform all department heads of the tentative schedule as soon as it is prepared so they can make appropriate plans. Based on information supplied by your mover, let the department heads know what is to be expected of their department personnel in preparing and/or packing

materials, supplies and furniture. As the moving date approaches, try to establish a detailed moving schedule down to the quarter or half hour. Keep all parties informed of this schedule, but be prepared for the inevitable unforeseen event that will disrupt it. Be prepared to make on-the-spot changes to adjust for the unexpected.

The schedule developed should try to accommodate people's wishes. For example, the moving company may want to load and unload a floor at a time, and the company might need to re-locate departments onto different floors and into different adjacencies in the new facility.

3. The movers will need floor plans indicating where the departments and individuals will be located in the new space. The movers should explain in detail how to key the plans for their specific needs. These plans should be keyed to the room numbers actually in place. These numbers may be different from those assigned by the architect during the preparation of the various space planning and construction drawings.

When preparing the floor plans for the movers, prepare an inventory of everything on the floor, whether it is to be moved or not. The inventory should be keyed to the moving plan — precisely. If an item is not to be moved, the inventory should state that information.

4. In addition to the main company liaison or move coordinator, have each department head appoint a liaison to work with the movers and to be responsible for his or her department's clean-up, move-out and move-in. The departmental liaison personnel will be responsible for answering questions from both the movers and from their department's employees.

5. Labels should be prepared with great care, as any errors could prove costly and very frustrating. Again, the mover should give specific instructions on how he wants labels to be prepared. He is the one responsible for getting the material to the right places on the right floors. There are two specific points to keep in mind in connection with labeling.

First, be careful if or when you are using tags tied to furniture or equipment. Tags may be torn off before the move is complete.

Second, some pressure-sensitive labels are easy to apply— but almost impossible to remove if left on for very long. Remove all labels as soon as they have served their purpose. Be sure to have a good supply of adhesive or gum remover that will not damage furniture finishes on hand.

6. Since most moves are charged for by the pound, be sure you only move what is needed. To help ensure an efficient and economical move, ask employees to do a thorough pre-move housecleaning. Urge them to get rid of all unnecessary files, paper and other items that have been accumulating over the years.

7. Before you move in, inform the landlord or contractor of a moving date, review all the construction work to be sure that it has been completed properly, and that construction materials will not obstruct the move. Having painters or electricians trying to work while a move is in process can not only cause disruptions, but create excessive delays both in construction and in the move. Walk the new space. If you see scratches, marks on walls or doors, soil spots on carpeting, etc., make a note of each item. Bring the items to the attention of the landlord or contractor so the damages will not be blamed on the mover.

Walk both the old and the new space with a representative of the mover. Invite him to prepare his own list of existing conditions. Come to an agreement as to the pre-move conditions. You will wish to hold him responsible for damage to the old and the new facilities as a result of the move. The pre-move condition examinations and lists will be valuable in fixing responsibility for damage and in settling claims.

You will also want to examine all newly purchased furniture and furnishings for correctness and proper condition at the time of delivery. Information on how to do this can be found in the discussion on punch lists in Chapter Eight. Follow the same procedure for furniture punch lists as you did for construction. As with the new and old space conditions, make a list of all obviously damaged furniture or equipment which is to be moved *before* it is moved. Invite the moving company's representative to make his or her own lists of any observed damages. This list will go a long way towards arriving at an evenhanded adjustment of damage claims which are likely to arise from the move.

8. Several days before the move, give each employee a detailed moving check list, which should include instructions from the mover on what each individual is to pack and how items should be labeled. There should be specific information as to the floor to which the employees should report in the new facility, where they are to park, and any other information they might need to aid in the move. Small maps and floor plans are helpful.

9. The move coordinator should prepare a master list of each employee, workstation, and areas such as storage rooms, library, conference rooms, etc., in the new facility. Each should be assigned its own identification number and color code. Depending upon the complexity of the move, this coding system may designate to which floor and to which part of a floor the various items are to be moved. Furniture, containers and other items should not be moved by employee name.

10. Once the move has begun, you should delegate one or two people to keep account of the mover and his crew. This is particularly important if you are in a state requiring a time and material contract. How many trucks are involved? How many drivers, loaders and unloaders, elevator operators, floor people, packers, foremen and supervisors are there on the job and where are they working? Were they there all day? Arrange for the mover to provide you with copies of time sheets, weight tickets from a certified weighing station, and dispatch tickets or driver's logs indicating which trucks arrived on the job at what times, left the job at what times, arrived at the weighing station then at the new location, at what times. As quickly as possible verify the movers logs, tickets and time sheets with the information prepared by your crew. If discrepancies appear, have a meeting with your crew members and the moving company representatives – before the invoice is received. Try to settle any apparent differences while information is fresh and before positions harden and are memorialized in the form of invoices.

11. Figure 9.2 is a sample set of instructions to be given to each employee to assist in preparing for the move. When each employee is given an instruction sheet, it should contain his or her unique moving number and color code. This form can be found in the Master Forms Section.

12. Request that the mover have a small crew available to locate each piece of furniture or equipment to its exact new location after it is moved from the elevator unloading area. The crews working between the elevator and the floor are under pressure to clear the elevator area as quickly as possible. As a result, they will not be able to spend time positioning each piece of furniture or equipment to your exact needs. In the same regard, be sure the person you have delegated to be responsible for his or her department or group during the move is on hand to provide exact direction to the movers.

Facility: _____   Location: _____

Employee: _____   Department: _____

Floor: _____ Room: _____   Number: _____ Label Color: _____

**YOUR MOVE WILL BE HANDLED MOST EFFICIENTLY
IF YOU CAREFULLY FOLLOW THESE INSTRUCTIONS**

Placement of your furniture at the new location is done by number - NOT BY NAME.

Your moving number and label color are listed above. Items without a label will not be moved. You will be supplied with labels and tags.

Marking: Label each item. If an item must be dismantled to be moved, be sure to label ALL parts. Example: A secretary desk will receive one label. The desk's typing return will receive a separate label.

Desks: Label the desk on the top surface.
- Your desk will probably be turned on end. Pack accordingly.
- You will be given a moving container(s) for loose items. Pack it carefully.
- Place small items such as pen, clips, rubber-bands, etc. in an envelope in your center drawer, or pack in your container.
- Breakable items and liquids such as ink, glue, etc. should be removed and packed in your container.
- If your desk is to be carried up or down stairs, empty it completely.

Glass tops on desks: Place label on lower right-hand corner and remove all papers from under the glass.

Filing cabinets: Place label on top drawer. Be sure all pressure plates are moved forward so contents will be secured. It is not necessary to lock drawers.

Bookcases: Place label on top or on sides. Contents should be removed and packed in containers.

Large Supply Cabinets: Place label on front near upper right-hand corner. Remove contents and pack in containers. Label these containers the same as the Supply Cabinet.

Typewriters, Word Processors, PC's, and other machines: A special rack is used to move office machines.
- Secure typewriter carriage by placing both margin stops in the center. Place label on platen. Pack the cover in your container.
- Pack PC's and processors according to manufacturer's instructions. Lock all disk drives. BACKUP ALL FILES.

Figure 9.2

- For all machines, place label on the top or on front.
- Unplug all cords, pack in your container if removable. If not, wrap cord around machine (or lamp).

Packing Containers: Place label on BOTH ENDS. Not on top. Containers are stacked when moved. Please do not over-pack. Get another container if necessary. Flat tops speed the move.

Miscellaneous:
- Do not forget to tag loose seat cushions and backs, label carpet pads, wastebaskets, etc.
- Labels are usually easily removed when the move is completed. Sometimes they will not adhere to some surfaces. Try placing a strip of scotchtape over the label or use a tag.
- The mover will not be allowed to remove items attached to the walls. Detach and label if they are to be moved.
- If you have questions contact your coordinator.

Name: _____

Present Extension: _____   New Extension: _____

Figure 9.2 *(continued)*

13. Once the move has been completed, be sure the mover follows up on any necessary adjustments or last minute changes, and promptly collects all of his moving equipment, including containers.

14. The success of the move will be in direct relation to the degree of planning and control you and the mover have employed.

Orienting Employees to the New Facility

Part of the successful accommodation of the company's employees to the new facility will be the result of how well prepared they are for their new environment and equipment. Hold at least one orientation meeting before the actual move-in to describe the new facility and its location. Explain the facility design, the new standards and furnishings. Give special attention to any major changes of departmental adjacencies or changes in organizational structure. If such a meeting is not possible, give each employee a printed handout similar to the example in Figure 9.3 Add your own specifics about parking, public transportation, food service facilities – both in-house and public, safety concerns if any, operating hours, visitors, and any other pieces of information relevant to the new facility. It is helpful to include area-wide and local maps, together with any other information which may assist the staff in making a quick and comfortable adjustment to their new surroundings.

WELCOME TO OUR NEW OFFICES

Thank you for your help and cooperation in the recent move to our new offices. They are the result of much thought, planning and work towards the goal of improving working conditions for us all.

In this memo we would like to tell you about some of the important features we have incorporated into the new offices, and to establish several new housekeeping procedures.

Furniture

Your workstation has been custom-designed for a person in your job category. After analyzing your work needs, we designed a system of components which provided work surfaces, tack surfaces, chairs, files, and storage spaces needed to help you in your work. If you feel you need additional or different equipment, please contact the Move Coordinator:

Name: _____ Extension _____

so that your request may be evaluated.

Your workstation is of high quality. Please do not attach anything directly to it. If you wish to post something within your workstation, use the tack surface. If you need to attach additional materials, please contact your Move Coordinator for the proper way to do so.

Pull-up guest chairs have been placed near some workstation areas for those times when additional seating is required. Please return them to their original locations after they are used. Do not use chairs as additional storage surfaces or your private coat rack. The chairs are for people to sit on.

If you have any questions or problems about the placement of furniture within your area, please contact your Move Coordinator. No changes or additions may be made without approval.

Desk Accessories

We have supplied basic desk accessories and a wastebasket for each workstation. To keep desk tops as neat as possible and to assist the daily cleaning crew, do not leave papers out at the close of the day. Place files and outstanding work inside your storage units or file cabinet or return them to Central Files.

Consult your Move Coordinator before cleaning the top of your work surface. Specific cleaning products have been suggested by the manufacturer. Using anything else may damage the furniture finish.

Counter Tops

Keep tops of all counters, files, and storage units throughout the facility free of miscellaneous papers and other items. (Note, delete the following if not applicable to your facility.) Plants will soon be placed on some of these tops in accordance with our design plans.

Integrated Lighting System (If applicable)

To produce a pleasant working environment and to conserve energy, a lighting system has been integrated within each workstation. The correct light level is supplied to your work surface, and at the same time a lower-than-usual ambient light is provided overall. This system is designed for its efficiency and quality of light. However, we recognize that not everyone requires the some light level for their work. If you feel the need for a modification to your lighting please contact your Move Coordinator for assistance.

Nothing is to be put on the walls or doors. If you are in a private office and feel that you need to post charts or other material, please contact the Move Coordinator for the proper way to do so.

Signage

If this has not already been done in your area, directional and identifying signs will soon be put up. In the meantime, please do not post temporary signs.

Artwork (If applicable)

We have selected art to be hung and placed in general use areas. No additional artwork is to be put up without special permission from your supervisor. No miscellaneous photographs, posters, clippings or other items may be put on either walls or furniture. Personal photographs may be attached to your tack surfaces or placed on your work surfaces.

Plants (If applicable)

Plants, which are to be maintained only by the supplier, will be placed throughout the facility. Please observe the supplier's request that they are to be the only people watering or feeding the plants.

Figure 9.3

No plants or floral arrangements may be added without special approval, except those delivered to the office on birthdays, anniversaries or in celebration of other special events. These may remain on your work surfaces.

Heating and Air-Conditioning

Although we do not anticipate any problems with the air-conditioning, it make take a few attempts before it is adjusted to everyone's satisfaction. If you find the temperature in your area to be consistently uncomfortable please notify your Move Coordinator.

Please take a little time to become familiar with our new space. If you have suggestions for improvements, please send a memo to your Move Coordinator for review with management.

We are delighted with our new offices and hope you are too. We have made every effort to create a pleasant working environment for you. While this new space may differ from our old office, once you begin to work in it you will discover some of its benefits, including better work flow, better storage, less noise, more privacy and more colorful furnishings. We believe you will find these new offices comfortable, efficient and a pleasant place to work.

Following is a list of people you should contact for the following:

Telephone problems:

Name _____ Extension _____

Computer problems:

Name _____ Extension _____

Broken or damaged furniture:

Name _____ Extension _____

Furniture adjustments:

Name _____ Extension _____

Figure 9.3 (continued)

Again, this is only a suggested memo. Adjust it to suit your situation and office policies.

In addition to giving each of the employees the above memo, provide them with data about the area into which they will be relocating (or have been relocated). Provide useful information on finding the day-to-day necessities both inside and outside the facility.

It might be beneficial to have an employee open house to which employees and perhaps their families might be invited. This can have a couple of benefits, first to help orient employees to their new surroundings in an informal atmosphere; and second, it is a good morale booster after what may have been a difficult and disruptive relocation experience. Some companies will provide welcome packets or baskets that contain local information and or goodies.

The orientation information should be distributed as a part of the welcome packet to the employees before the move.

Suggested orientation information includes:
- Where the new building is located and where the entrances are located.
- Routes to and from expressways and major streets
  (If possible include NO TURN and ONE WAY street designations.)
- Bus or other public transportation routes, designated by bus or train numbers and the location of loading and unloading stops. Schedules can be obtained from your public transportation authority.
- Parking lots, if related to the building. If not, where there are areas of commercial parking.
- Gasoline service stations and auto repair facilities.
- Banks and ATMs.
- Restaurants, fast food, convenience markets, drug stores, etc.
- Dry cleaners, laundries and other sundry services.
- Child care centers.
- Hospital and emergency facilities.

Observe how the employees are adapting to their new workstations:
- Are they using the components properly?
- Do they appear comfortable, particularly in their posture and leg room?
- Are work surfaces and seating at proper heights for each individual?
- Are lighting, noise and temperature levels appropriate?
- Are there sun glare problems? These usually occur early and late in the day.
- Is circulation working out as planned? Are there congestion points? Are some areaways too wide or too narrow or too long?
- Keep trying to adjust or balance the HVAC system to accommodate the greatest number of people possible. You will not succeed in satisfying everyone all of the time. Remember, the 74°F which seemed comfortable to you yesterday may seem too hot or too cold today.
- You will take some heat, but the employees will appreciate your interest and concern — and you might learn something valuable for the next installation.

Follow-Up

Relocating the company's office will be a somewhat traumatic event for everyone. Not all employees will be enthusiastic nor will they appreciate the effort which has gone into the planning and execution of the project.

Some people will be outspoken in their criticisms of the location, the workstations, the color schemes, the air temperature, the whatever. A good bit of this is nothing more than a lack of adjustment to the new environment. However, some portion of these gripes will be well founded.

Learn to listen to these gripes. While assisting with the new building, you have been working with broad principles, most of which are accurate most of the time. But nothing and no one is perfect. Within these criticisms may be some very valuable information that could help facilitate job performance. Delay some changes for a period of perhaps 90 days to allow people to adjust to their new surroundings. Many gripes may disappear during that time.

The facilities manager, the move coordinator and the project manager should maintain a high profile during and after the move. A good sixty days is not too long. Adopt the adage "See and be seen."

Chapter 10

FOLLOWING UP AFTER OCCUPANCY

Introduction

At the end of the first week after the company's move, begin a series of reviews to see if all equipment is being used correctly and is functioning properly, and if the employees are satisfied with their new furnishings. The pre-move-in review and punch lists were completed to be sure all installations coincided with the furniture plans. The reviews in this chapter will help confirm that the facility really works.

Reviewing the Facility in Operation

Walk through the facility, area by area. Pay particular attention to overall noise and light levels. Check carefully to see whether each piece of equipment is being used as expected. In most cases the problems found will be minor and easily corrected by simply instructing the employee on the correct procedures. Here are some other things to watch out for:

- The paper organizer: does the person using it understand how it functions?
- Are files stacked on the floor rather than in file cabinets? Is there enough file storage space?
- Are there inactive files which should be sent to storage?
- Are electrical, phone and data lines run through the workstations properly, or are they draped across the work surface or on the floor?
- Are work surface returns, video display terminals and keyboards, shelves, and other components at the proper heights?
- Are there glare problems from lighting sources or from windows? Should glare screens be installed? Should additional lighting fixtures be installed?
- Is there adequate room for people's legs between pedestals and under work surfaces?
- Are there any sharp edges or protruding fasteners such as screws or staples damaging hosiery or clothing?
- Are file drawers in the correct position? Do they open and close easily and stay in position?
- Is all equipment properly leveled and stable?
- Are there traffic congestion points that may necessitate a change in furniture arrangement?
- Are the graphics and signage clear and understandable?
- Has any furniture, such as a table, file cabinet, or side chair been put in the wrong place?

- Should plants be re-grouped to break up large empty areas, or to open up congested spaces?
- Is the artwork in the correct locations? Should it be repositioned to break up wall spaces or to provide visual relief?
- Is the lighting on stairs bright enough so that employees will not hesitate to use them?
- Are chairs adjusted to comfortable heights, and for proper swing tension in back or tilt? Do all employees know how to adjust their seating?
- Is the overall lighting adequate? Is additional ambient or task lighting required?
- Is the heating and cooling reasonably comfortable?
- Is there any way the facility can be adjusted to better meet employee needs and job requirements? From a practical and legal point of view, is the Americans with Disabilities Act being met?

During the review process places may be found where additional equipment is required, or where too many pieces have been ordered and are not being fully or properly used, such as:

- Drawer pedestals
- Bookshelves
- Paper organizers
- Side chairs
- File cabinets
- Panels
- Task lighting
- Desktop accessories
- Coat racks
- Word processors and typewriters
- Adding machines or calculators
- PC and printer stations
- PC keyboard drawers or trays
- Fax stations

Make a list of any such items.

Talk with the employees. Encourage them to submit any problems, complaints, requests, or suggestions to their department head or supervisor. In turn, the department head or supervisor should compile a list of these comments for the facilities manager or coordinator to review.

Ask the person compiling the list to note when and how often a particular complaint or request is repeated, in order to give you some idea of whether it is a major problem or simply an individual one.

It may happen that the workstation selected for a particular job function is not working out as anticipated. Correct it. Tell the employee it will be corrected. Remember, acquiring alternate components to effect the correction may take weeks; keep the employee informed.

After your lists and those compiled by the department heads and supervisors have been reviewed, the coordinator will supervise any necessary corrective steps.

If additional items must be purchased, the coordinator should issue a purchase order, preferably to the same supplier from whom the initial quantities of the item were purchased.

If any changes require the services of the contractor, such as moving a light fixture, or reversing the swing on a door, the coordinator will issue a change order, following the previously outlined procedures.

As another way of reviewing the facility in operation, utilize the suggested questionnaire (Figure 10.1). *Evaluating Our New Offices*, or the company's own questionnaire, should be filled out by all employees.

Distribute this form one to two months after the move to the new space. Waiting a couple of months lets people become more accustomed to their new surroundings, and real problems will become better defined. Immediately after moving in, advise all employees that shortly they will be asked to complete a questionnaire with their comments. A full-sized copy for reproduction is in the Master Forms Section.

Maintenance Instructions

Creating the new facility has been a long, hard job. To keep it in top condition, have all employees follow standard maintenance procedures, especially those given by the manufacturers of the major items purchased.

After reviewing the maintenance procedures, set up a schedule for regular cleaning. (This schedule should be reviewed with the Building Manager or landlord.)

Final Cost Settlements

Be sure to make prompt payment on all accounts related to the project. However, if invoices are received for progress or final payments for work which has not been satisfactorily completed, withhold a portion of the payment until matters are properly settled. In the construction industry, it is normal to withhold ten percent of each progress payment until all work involved in the contract is completed. When purchasing furniture, equipment and the like, it is normal to withhold part of the final payment to cover the cost of incomplete or damaged work. Pay the withheld funds when repairs or final delivery have been made.

Do not withhold the entire payment from a vendor if the major portion of his work has been completed satisfactorily. Try to determine a reasonable amount to be withheld, one which is appropriate to the amount of incomplete or unsatisfactory work. Do not make final payment until all invoiced work has been completed to company requirements.

Upon completion of the project and payment of all invoices, issue a final budget report. Do this by typing the words "Final Budget Report" at the top of the *final budget form*. Because of changes which may have occurred during the course of the project, and because some items may have been purchased on a time-and-materials or open contract basis, the final figures probably will not exactly equal the previous revised total figure.

Distribute the final budget report to the coordinator, the branch manager, and to the officer in charge of the facility. Review it to determine where the greatest errors or omissions occurred. Consider how the project could have been made more successful in terms of facilities function, cost and scheduling.

Facility: _____ Location: _____

Employee: _____ Department: _____

Phone: _____ Date: _____

This questionnaire is being distributed to all employees to help us evaluate our new office space. Your answers will help us not only to learn how well our new facility is working, but also will aid us in planning future facilities. We would appreciate your answering the questions as honestly and fully as possible. Although we would like you to sign your name and indicate your department in order to follow up on problems, you do not have to do so. Thank you.

What do you feel are the most noticeable differences between the new facility and the old one?

Good: _____

_____

Bad:_____

_____

What aspects of the new offices are disagreeable to you (if any)?

_____

_____

Do you personally find the new environment conducive to more efficient work habits?

If yes, why? _____

_____

If no, why? _____

_____

The noise level is: pleasing [ ] displeasing [ ] (check one)

_____

_____

_____

Figure 10.1

The colors used throughout the space are:
pleasing [　] displeasing [　] (check one)

In what ways. Please specify: _____

_____

_____

You have:
not enough room [　] enough room [　] too much room [　]
in your new workstation or your new offices. (check one)

What do you like about your workstation? _____

_____

What do you dislike about your workstation? _____

_____

What would you suggest be done to improve your workstation? _____

_____

Do you feel the various departments are efficiently laid out from an organizational and working standpoint? Please comment. _____

_____

Do you think the arrangement could be improved? If so, how? _____

_____

Do you feel the new environment has: improved [　] or harmed [　] work attitudes of your co-workers? (check one).

Please explain. _____

_____

Your group performs:
more work [　] the same amount of work [　] less work [　] since the change to the new offices. (check one)

Comments. _____

_____

_____

Figure 10.1 (*continued*)

Do you feel that the change has been for the better or has made your working conditions worse? Please explain how.

_____

_____

_____

Are there any modifications you would like to see to improve working relationships within your department? Please explain how.

_____

_____

_____

General comments:

_____

_____

_____

_____

_____

_____

_____

_____

_____

_____

_____

_____

_____

_____

Figure 10.1

Part II

TECHNICAL SECTIONS

The following chapters deal with specialty aspects of the facility. These include:

- Floor Coverings
- Furniture and Furnishings
- Accessories, Graphics and Signage
- Plants and Containers
- Art
- Lighting
- Acoustics
- Color, Paint and Wall Coverings

Information on these subjects has been included because these components or conditions require judgment calls and careful selections. The space planner, designer or architect may not be able to provide assistance in making qualified selections. Many consultants may have a good knowledge of some aspects of carpet construction or color selection, etc., but usually they must depend on some other third party for technical knowledge. It is important that space plans, design, specified components, furnishings and materials be attractive and functional. It is also important that they properly serve their anticipated function at an economical price.

The following chapters provide information on material composition and construction, fire- and smoke-related codes, product availability, lead times, selling and distribution practices, and the physics, physiology and psychology of certain ambient conditions.

These chapters should be reviewed when constructing the specifications or purchasing a system or product described.

An attempt has been made to provide information on the major systems and components which are likely to concern a facility or project manager. A section on heating, ventilating and air-conditioning is not included. The HVAC systems operating in nearly all buildings today, and those available in the near future, are ambient-oriented. Most of these systems supply temperature-controlled air from diffusers, or from cabinet units into pre-selected, fairly large space volumes. The input air temperatures of these large space volumes are controlled by local thermostats. While these systems provide a fairly comfortable environment, they cannot satisfy all occupants all of the time.

Systems designed to provide more localized and individually controlled heated and cooled air are becoming available. Some work has been done on integrating HVAC systems into individual workstations. Systems are being developed that utilize floor distribution so that temperature-controlled air can be provided on a custom basis to each individual workstation. Highly sophisticated sensory devices are being developed to measure temperature and humidity and to provide information to controls, which are quicker in their responses to changing conditions.

For the present, the company's mechanical engineer or the landlord's mechanical engineer should be able to provide reasonably comfortable temperature and humidity conditions on an ambient basis.

Chapter 11

FLOOR COVERINGS

Introduction

There are approximately six varieties of floor coverings appropriate for office use. Of these, carpeting is the most widely used and specified. Carpeting accounts for nearly 90 percent of the floor covering used in most office facilities today. The next most frequent is composition floorings such as vinyl composition tile (VCT) and rubber tile. The remaining materials are wood, ceramic tile (including quarry tile), terrazzo and, lastly, natural stone such as marble and granite. There are any number of variations and combinations of the basic materials. Each has a unique niche in the design process.

Quarry tile is normally the best material to use in a food preparation area such as a cafeteria kitchen. Unglazed ceramic tile is the best material for restroom floors, as it is basically non-slip and impervious to most acids. Rubber flooring is superb in areas where people must stand for long periods or on interior ramps, such as a ramp up to a raised computer room floor. Terrazzo is excellent in very high-traffic areas subject to heavy soiling, such as main corridors on grade level. Wood can add a stylish touch to an executive suite, and natural stone can give a main building entry a classical feel.

However, carpeting, by its nature, is the most widely used floor covering in most facilities. This chapter does not attempt to provide a complete technical background in carpeting, but does provide enough information for informed product selection.

Carpet, wall or *cove base* is typically supplied and installed by the carpet installer and should be a part of the installation price. It is typically supplied 1/8" thick, in 2-1/2", 4" or 6" heights, and 48" lengths. Carpet base is formed as essentially a flat material with a slight return at the top edge for a snug wall fit and some grooves in the back surface for mastic retainage. Cove base is typically formed the same as carpet base and is supplied in the same dimensions. Cove base has a down-angled, flexible toe shaped into the bottom edge, for a snug floor fit. Both carpet base and cove base are available in a wide variety of colors.

Wall base must comply with the following code requirements:

Federal Specification SS-W-40a, Type I, and ASTM designation E-84, "Tunnel Test".

As for the carpet itself, this chapter covers:

- Fibers
- Yarns
- Tufting
- Dimensions
- Dyeing
- Testing
- Maintenance
- Specifying
- Installation

Fibers

The vast majority of carpet fibers used today are man-made. Only occasionally are wool, cotton carpeting or rugs being used. Their application is so unique that they are not addressed here, particularly as they cannot pass any of the fire rating tests.

Acrylic

Produced by Monsanto Textile and BASF; known as *Acrilan*, *Orlon*, and *Zefran*. Available as level loop, cut pile, and velva-loop. This fiber is practically nonexistent as a carpet material.

Ratings

• Moisture absorption	To 1.5%
• Effects of heat	Melts at 420-490°F
• Burn Test I.D.	Forms black bead
• Dyes Used	Acid-cationic, some solution-dyed

Resistance to:

• Sun	Outstanding
• Mildew	Outstanding
• Aging	Outstanding
• Stain/soil	Good
• Acid	Good
• Alkalis	Fair to good
• Solvents	Good

Characteristics

• Resilience	Good
• Abrasion	Good
• Color Retention	Good

Nylon

Produced by Allied, DuPont, American Enka, Monsanto Textile, BASF. Known as *Anso*, *Antron*, *Camalon*, *Enka*, *Enkalon*, *Ultron*, and *Zeftron*.

Available as level loop, cut pile, high-low loop, velva-loop and saxony. A major fiber in commercial carpet manufacturing.

Ratings

• Moisture absorption	To 4.5%
• Effects of Heat	Melts at 320-375°F
• Burn Test I.D.	Shrinks, difficult to ignite
• Dyes Used	Acid-cationic, some solution-dyed and differential

Resistance to:

- Sun — Some degradation
- Mildew — Excellent
- Aging — Excellent
- Stain/soil — Good
- Acid — Fair to good
- Alkalis — Superb. Soluble in formic and phenolic acids

Characteristics

- Resilience — Good
- Abrasion — Superb
- Color Retention — Very good

Olefin

(Polypropylene) Produced by Hercules, Amoco
Fabrics, Chevron Chemical, Phillips Fibers, Strudex Canada. Known as
Herculon, Marquesa Lana, Marvess, Patlon, Polyloom, Strudon.

Available as level loop, cut pile, velva-loop, and saxony.

A major fiber in commercial carpet manufacturing.

Ratings

- Moisture absorption — To .01%
- Effects of heat — Melts at 250-333°F
- Burn test I.D. — Shrinks, turns to light brown hard bead, celery smell
- Dyes used — Colored at time of extrusion, some post-extrusion dyeing

Resistance to:

- Sun — Fair to good
- Mildew — Excellent
- Aging — Good
- Stain/soil — Very good
- Acid — Excellent to most acids
- Alkalis — Excellent to most alkalis
- Solvents — Caution; some solubility

Characteristics

- Resilience — Fair
- Abrasion — Superb
- Color retention — Excellent

Polyester

Known as *Dacron, Fortel, Kodel*

Available as cut pile, high-low loop, velva-loop, and saxony. This fiber is primarily used in residential carpeting.

Ratings

- Moisture absorption — To .04% to .08%
- Effects of heat — Melts at 445-455°F
- Burn test I.D. — Shrinks to hard bead, pungent smell
- Dyes used — Cationic-disperse and continuous

Resistance to:

• Sun	Some strength loss under prolonged sunlight
• Mildew	Excellent
• Aging	Excellent
• Stain/soil	Good
• Acid	Excellent to most acids
• Alkalis	Good to most alkalis
• Solvents	Good; some solubility

Characteristics

• Resilience	Good
• Abrasion	Good
• Color retention	Good

Yarns

The techniques for producing yarns and plies are varied and somewhat complex. We will deal only with man-made fiber.

The initial phase of producing yarn begins with the extrusion and drawing of the individual fiber. Various plastic granules are mixed to develop the particular fiber structure desired by the manufacturer. Often dye chips are added to the dry, granule mix to produce a colored fiber (solution dyeing). The dry granular mixture is heated and extruded through small holes in a metal plate. The extruded fiber is cooled and wound onto a creel. The extrusion process gives the fiber permanent and special characteristics, such as a cross-sectional shape, size, and special dyeing properties.

The fibers are pulled through a stretching process and then through a *bulking* or *crimping* process where they are given a wavy character. This process gives the yarn a less prickly, softer feel.

At this point, some fiber may be ready to be processed into carpeting. The fiber is cut into staple lengths of 3-1/2" to 8" and baled. These staples are blended with other staples, combed to produce selected ropes of yarn called *silvers*, which are twisted together to be spun into a yarn, either before or after dyeing. The yarns are twisted and heat-set. The resulting yarn can then be tufted into carpet.

Yarn Size, Denier, Cotton Count

There are two systems of describing yarn size — denier and cotton count.

Denier:

The denier, or so-called direct method, is based on the metric system. It measures the weight in grams of 9,000 lineal meters (1 meter equals 39.37 inches) of yarn. If, for example, 9,000 meters weighs 1,800 grams, the denier is 1,800. Expressed 1,800/1 for one-ply, or expressed 900/2, for two-ply. *Denier* is generally the term used with Nylon and Olefin yarns, which have continuous filament structure.

Cotton Count:

Cotton count is opposite in concept and is an indirect method. It is the number of lengths of yarn required to weigh one pound. Cotton count is the number of 840-yard lengths needed to weigh one pound. If it takes two lengths to yield one pound, the CC is 2.00/1 for one-ply, or 4.00/2 for two-ply. Cotton count normally is used in dealing with cottons, wools, acrylics and polyesters, in other words, with staples.

Tufting

This section deals with some of the terminology, processes and specifications of tufting. Woven carpeting is also covered in this section.

Woven

Woven carpeting has a pile, the loops of which enter the backing through one hole and return through a different hole, much like needlepoint. While woven carpeting is still readily available and is considered to be the more durable in terms of construction, it is rapidly being replaced by the tufted type which accounts for over 95 percent of commercial carpeting. This change is due, in large part, to the highly improved technology in backings.

Tufted

The pile in tufted carpeting consists of yarn loops which enter and return through the same hole. The loop can be left as a loop or cut to form a cut pile, velva-loop or tip-sheared. The initial production product consists of the primary backing, through which the tufts have been set. It is called *greige goods* and is extremely flimsy. A secondary backing must be applied to the greige goods.

Tufted carpeting is now a near equal, in strength, to woven carpeting. Tufted carpeting's greatest weakness was snagging and pulling. If a loop was snagged and pulled, a long string of yarn would be pulled from the backing, leaving a long gap in the carpet face. However, current backing technologies have provided much stronger locking characteristics that now render tufted carpet much more resistant to *pulls*.

There are two basic types of backings. One type uses a secondary fabric backing of jute or polypropylene, which is laminated to the stretched greige goods with a latex binder. The two fabrics are rolled together with the latex and then passed through an oven where the latex is cured. Carpeting that is to be installed over padding and tacked down using a carpet tack-strip should have a secondary fabric backing.

The second type of backing is called *unitary backing*. Unitary backing consists of applying a heavy, highly penetrating latex, often foamed, to the stretched greige goods. The back-coated goods are fed into an oven where the latex is cured. Carpeting that is to be installed using a direct glue-down method typically uses a unitary backing.

Dimensions

There are three basic dimensions used in carpet construction – gauge, pitch and weight.

Gauge

This is the number of needles (tufts) per lineal inch counted across the width of the carpet and stated in fractions of an inch. The dimensions are taken from the center of one tuft to the center of the adjacent tuft. The numbers can be expressed as follows:

Gauge Width	Needles/Inch	Needles/12 ft.
5/32	6.4 (32/5)	922 (6.4x12"x12')
1/8	8	152
1/10	10	440

Pitch

This is an alternate method of describing gauge. It is expressed in the number of needles per 27 inches of carpet width. It was commonly used prior to the development of *broadloom* — rolls of six feet wide or greater. The term broadloom is fairly well out of date.

SPI — Stitches Per Inch

This is the number of needles (tufts) per lineal inch counted the length of the carpet. Unless there is a compelling reason to deviate from the norm, the SPI is usually the same as the gauge.

Height / Pile Height

Pile height is measured from the base of the tuft, at the upper surface of the primary backing surface, to the top of the tuft. If there are two levels of tuft, the average height is used. It is expressed in fractions of an inch or in a decimal equivalent.

Weight

One of the most common terms used in the carpet industry is *weight*. Weight is expressed two ways, *face* or *pile weight* and *total weight*. Total weight is a pretty useless measurement in determining the quality of carpet, as it consists of both the yarn, the primary backing, the latex and any secondary backing.

Face Weight

Face weight is the more common term used and is also a fairly confusing term. Face weight consists of the yarn weight only per square yard, expressed in ounces.

The terms which provide the most complete information as to relative quality are *pile density* or *weight density*.

Pile Density

Pile density is the result of multiplying the finished pile weight x 36, divided by the finished pile height.

EXAMPLE:

$$\frac{36 \text{ x } 32oz/sy.}{.25 \ (1/4")} = 4608$$

Weight Density

Weight density is the result of multiplying pile density x face weight (in ounces). This is essentially a Federal specification and not often used other than for federally specified work.

Dyeing

There are about seven basic dyeing processes.

Solution or Pigment Dyeing

Taken in sequence of when they occur in the manufacturing process, the first would be *solution* or *pigment dyeing*. In this method, the raw plastic granules are mixed with pigments before the plastic is melted, extruded and spun into a filament. The color is permanently a part of the fiber.

Solution dyeing is the most common method of dyeing for commercial-grade carpeting. It is the most resistant to staining and fading, especially if treated with an ultraviolet light inhibitor.

Stock Dyeing

The next is *stock dyeing*, in which the fibers (while in staple form) are placed in a vat of dye. This process assures a highly consistent color.

Space Dyeing

Space dyeing applies the dye to yarn which is constructed of specially treated fibers which will take one or more colors of dye along its length. The result is a tweedy looking carpet after tufting.

Skein Dyeing

Skein dyeing is a vat dyeing process in which the yarns are removed from their creels, bundled into a bale and dyed in a vat of very hot dye. Smaller quantities of special colors can be achieved this way at economical cost.

Piece Dyeing

Piece dyeing is a completely tufted greige which is immersed in a dye vat, dried and backed.

Differential and Cationic Dyeing

Differential and *cationic dyeing* are similar to piece dyeing, the difference being that the yarns (fibers) have been specially treated along their lengths with *dye sites* that react to the dye differently and produce different shades of the same color. The difference between differential and cationic is the type of dye sites formed into the surface of the fiber.

Print Dyeing

Print dyeing is somewhat like silk screen printing. Dyes are squeezed or injected through patterns or by numeric-controlled needles directly onto the finished carpeting. The dyes can produce highly complex patterns, some of which may be custom designed. The costs and speed of production, and therefore lead times, are all quite good.

Testing

This section is concerned with only two types of test data—fire/smoke and static/electrical resistance. There are many other tests which have been developed: tuft binding, colorfastness, sound absorption, moths, delamination, breaking strength, etc. Most of these have been developed by the Federal government. Those who feel that any of these tests are important to their particular needs might wish to contact:

The Carpet and Rug Institute
P.O. Box 2048
Dalton, Georgia 30722
Phone: 706/278-3176

Fire and Smoke

A carpet selection should meet these basic fire and smoke tests:

OC-FF1-70 (Methenamine Tablet Test) This is a **must pass** test.

ASTM E-648 and NFPA-253 (Radiant Floor Panel Test)
This is a graded test.
Class I (.45 watts/sq.cm.) is required for all office use except where fire sprinklers are installed, then Class II (.22 watts/sq.cm) is acceptable.

ASTM E662, NFPA 258 (Smoke Density Test)

ASTM E-84 or UL 992 (Tunnel and Chamber Tests)
These tests are not often required, although some local fire departments may require them, especially if the carpeting is installed on ceiling or wall surfaces, such as an acoustical material.

Static and Electrical Resistance

There are three basic tests:

NFPA 56A IBM Method
> (Tests electrical resistance of material between an electrode and a ground. Measures carpet's ability to dissipate electricity.)

NFPA 56A Burroughs Method
> (This is a two-part test. The first is the same as the IBM test; the second measures the dissipation capacity of the carpet between two electrodes.)

AATCC-134 Electrostatic Propensity
> This is the most common test and criteria.
> (This test gauges the amount of static electricity generated by a test walking machine @ 70°F and 20% humidity.)

General Maintenance

Introduction

Carpet maintenance is a highly abused term. Maintenance assumes an ongoing process of preservation or upkeep. Yet most carpet maintenance is a series of corrective actions usually taken when the carpet condition has become an embarrassment or has gotten out of hand. The usual procedure is to call in the building cleaning company or carpet cleaning contractor and ask for a price to clean the carpeting. At this point, very likely, there is already trouble, and there could soon be more of the same.

Carpeting is a capital investment. It is the most utilized furnishing in a company's office. It is probably not possible to get to work without walking on the office carpeting. Most manufacturers warranty their carpeting for 10 years. As with most warranties, the warranty functions for the benefit of the manufacturer, and not the customer. It does this by pro-rating any claim and therefore limiting the liability. The manufacturer may claim improper maintenance. The manufacturer will most likely be right.

If, for example, the in-place cost of a selected carpeting is $25.00 per square yard, and the replacement cost is going up at the rate of 4% per year, the five-year replacement cost is in excess of $30.00, and $37.00 in ten years.

Improperly maintained carpet will often "ugly out" in about five to seven years. Assuming a required replacement at the end of seven years, three years have been lost from the capitalized life, or $7.50. Also, a replacement cost of nearly $33.00 has been incurred for a total of $40.50 per square yard, or $4.50 per square foot. A quality preventative maintenance program will prolong the attractive life of the carpet probably well beyond the ten-year warranty life.

Daily Cleaning

Good maintenance begins with good and regular vacuuming. Any area utilized on a daily basis should be vacuumed on a daily basis, slowly! Running a vacuum over a carpet at a breakneck pace does not do the job. Regular (daily) spotting will prevent stains from setting. There are a number of good multi-purpose cleaners, both water-based and oil-based, available, as well as a number of professional spotting kits. Regular power pile lifting should be employed. Finally, some form of regular surface cleaning system should be employed, either the dry extraction system described above or spray bonnet cleaning. Follow this with additional spot removal, as required.

Power Pile Lifting

This machine resembles and functions similarly to an upright vacuum sweeper. The differences are in the brushing and vacuum, which are both extremely powerful. Monthly to quarterly use of a pile lifter will keep the carpet pile in an upright condition and remove the dry dust and sandy soils from the pile and backing. Pile lifting will greatly improve shading in cut pile carpeting, but probably not eliminate it. Shading is a characteristic of all cut pile carpeting and therefore does not constitute a manufacturing defect. Unfortunately, many cleaning contractors do not even know that such a machine exists, although pile lifters have been around and in use for many years.

Spray Bonnet Cleaning

This process utilizes a standard rotary (semi-dry) buffing machine equipped with a fiber bonnet and a very fine sprayer for a nonresidual cleaner. After pile lifting, the carpet face is spray bonnet cleaned and spotted. The combination of pile lifting and deep vacuuming, together with the removal of surface oils and dirt, is extremely thorough and cost effective. The carpeting should be dry in about 30 minutes.

Programmed Maintenance

Programmed maintenance is eminently preferable to any of the aforementioned cleaning processes, with the exception of dry extraction, which could be considered a maintenance system if properly employed as such.

Setting up a maintenance program is a relatively simple procedure. Once in place, it should be reviewed a least once a year for effectiveness, and most certainly whenever there are substantial relocations or changes in office population.

The first step is to classify the concerned floor areas according to three categories:

- Light traffic: Less than 500 foot-falls in a 24-hour day.
- Moderate traffic: 500 to 1,000 foot-falls in a 24-hour day.
- Heavy traffic: Over 1,000 foot-falls in a 24-hour day.

In other words, if 200 people pass a particular point in a 24-hour day, the foot-fall count would be 400 foot-falls. If 500 people pass a point, the foot-falls would be 1,000.

Using a small drawing of each floor, color code the three respective traffic areas.

A reasonable maintenance program for each area might be:

- Light   Quarterly maintenance
- Moderate   Every other month
- Heavy   Once a month

Cleaning Systems

There are about five different carpet cleaning systems currently in use today. The differences between them, in terms of results, is significant.

Hand Scrubbing

This system is only rarely used today. There are a few contractors who still specialize in this process, which usually consists of a thorough vacuuming, followed by a hand scrubbing using a long-handled scrub brush dipped into a high-foaming detergent. This is followed by a vacuuming with a wet-vac. Spot removal should take place after scrubbing. Some contractors will try spot removal before scrubbing.

Advantages:

- Gentle to pile.
- Operator closely observes surfaces.

Disadvantages:

- Detergent is a residual type — leaving a deposit which attracts and holds dirt and oil.
- Wet-vac will remove 30% to 50% of the 1 to 2 gallons of water per 100 SF used in this system. In other words, leaving .70 to 1.4 Gal. per 100 SF.
- Slow process.
- Does a comparatively poor job.
- Expensive.

Reel Machine Scrubbing

This system employs a reel-type scrubbing machine which ejects a detergent directly onto the carpet while scrubbing. The reel machine's brushes rotate in a plane parallel with the carpet face. After scrubbing, the carpet is wet-vacuumed. Spot removal is usually done after scrubbing.

Advantages:

- Gentle to pile.
- Tends to lift pile into its original upright position.
- Does a comparatively fair job.
- Fairly fast.

Disadvantages:

- Detergent is a residual type (leaves a deposit).
- Wet-vac will remove 0% to 20% of the 1 gallon of water per 100 SF used in this system. In other words, it leaves 0.8 to 1.0 Gal. per 100 SF.

Rotary Machine (non-integrated type)

This system uses a rotary scrubbing machine injecting detergent directly onto the carpet, followed by a wet-vac extraction. Spot removal follows scrubbing.

Advantages:

- Does a comparatively good job.
- Quite fast (1,000 to 1,500 SF/man-hr.).

Disadvantages:

- Somewhat more abrasive to pile.
- Tends to lay pile over. (This is somewhat corrected by the wet-vac extraction and hand-brushing.)
- Detergent is a residual type (leaves a deposit).
- Wet-vac removes 0% to 10% of the .95 gallons of liquid per 100 SF used in this system. In other words, it leaves 0.85 to 0.95 Gal. per 100 SF.

Rotary Machine (integrated type)

This machine cleans exactly like the rotary machine previously described, the principal difference being in the integration of the wet-vac into the scrubber. This system removes 30% to 50% of the fluids, but only deposits about .75 Gal. per 100 SF.

Advantages:

- Does a very good job.
- Quite fast due to 1-step process (2,000 to 3,000 SF/man-hr.).
- Quite cost effective.
- Dries fast.

Disadvantages:

- Somewhat more abrasive to pile.
- Tends to lay pile over (This is somewhat corrected by the wet-vac extraction and hand-brushing).
- Detergent is a residual type (leaves some deposit).
- Integrated wet-vac removes 30% to 50% of the less than 1 gallon of water per 100 SF used in this system. In other words, it leaves 0.37 to 0.49 Gal. per 100 SF.

Hot Water (Steam) Extraction Method

This system uses a fine spray of hot water and detergent sprayed directly on and into the carpet. Some contractors claim that their system uses steam, although the steam has always looked like hot water to us. The fluids are left to soak into the carpet fibers for a few minutes. The extractor then wet-vacs the fluids out of the pile using either a drag wand or a light wand of 2 to 4 hp wet-vac.

Advantages:

- Somewhat more gentle to pile.
- Tends to keep pile in an upright condition.
- Does a comparatively good job.
- Quite fast.

Disadvantages:

- Detergent is a residual-type (leaves a deposit).
- Wet-vac extraction removes up to 85% of the 6 to 7.5 gallons of fluid per 100 SF used in this system. In other words, it leaves 1.0 to 1.4 gallons per 100 SF.
- Can cause glued-down carpet to release from floor and can cause some types of carpeting to stretch and form wrinkles. For these reasons, we do not recommend that this system be used without absolute guarantees from a highly professional, well-known and established service contractor.
- Dries slowly.

Dry Extraction

This system employs a four-step process. The first two involve the deposit of a cellulose material (often ground corn cob) saturated with a cleaning solvent onto a previously dampened carpet face. The third step involves the agitation of the granules on the carpet face using either a dry reel or a rotary-type brush. The final step is to vacuum up the now dirty granule material.

Advantages:

- May be somewhat more gentle to pile. (Questionable)
- Tends to keep pile in an upright condition if reel machine is used.
- Dries quite fast.

Disadvantages:

- Is apt to leave a dry residual of organic dust which has been ground into the pile.
- Not a deep cleaning process.
- Inconsistent spot removal.
- Fairly high cost.
- Slow (375 to 750 SF/man-hr.).

How to Specify Carpet

This section explains how to go about purchasing carpet, whether specifying and purchasing directly using a bidding process, or specifying and purchasing through the company's landlord as part of the lease agreement.

When Purchasing Carpet Directly

Any time the company purchases or contracts on a direct basis, it is undertaking responsibility and risk for the quality, cost and timing of the products or services purchased. The advantages can be a cost savings and possibly better control over the service. In the case of carpeting, the only choice being considered here is whether to have the landlord or his contractor purchase and install the company's carpeting, or have the company purchase and install directly. If the landlord does the purchasing and installation, the company may avoid some possible labor disputes (union vs. non-union). It may also avoid disputes over the condition of the floor surface prior to installation. The landlord or his contractor may have tighter control over tenant improvement schedules, and the company has less responsibility and risk of faulty installation or delays. However, the company may pay substantially more for the services of the landlord and his contractor. See the section, "If the Landlord Does the Work" in Chapter 2, "Evaluating Space Options."

When purchasing carpet directly, whether new or replacement, the company will first want to interview a number of prospective bidders. Select two or three and ask if they are interested in submitting bids, and what their general terms are likely to be.

Keep in mind the fact that you may be asking the bidder to install the carpeting in a union shop environment. However, do not assume that a union installation is mandatory. Discuss the likelihoods with each bidder and also with the landlord and the construction company representative.

Ask the bidders if they are willing to work on weekends and at what additional costs. Ask how they intend to protect the recently painted walls, woodwork, and any furniture which might be in place, as well as the carpet they are laying. Find out how much they think they can lay in a day. Are they familiar with the building? Either the company or the contractor will have to arrange for both dock time and elevator time.

Next, review and prepare the bidding documents by removing, copying, and completing the pages from the Master Forms Section as enumerated.

(Note: How much of the following document is used will, to some degree, depend on the amount of carpeting being purchased and its cost. Regardless of how much documentation the company prepares or how the company develops its bids, be sure to cover all of the points contained in these forms, in one way or another. Do not trust to the contractors' reputations. When it comes to a loss, the company's money or his, have no doubt about whose money the contractor is going to protect.)

Section A.   Instructions to Bidders
Fill in necessary information where indicated.

Section B.   Proposal Form and Instructions
Fill in necessary information where indicated. Include two copies of the proposal form, one for the contractor to keep, one for the company to keep after the contract has been awarded.

Section 1.   General Conditions
No fill-in necessary.

Section 2.   Special Conditions — Carpeting
Fill in necessary information where indicated.

Section 3.   Installation Specifications, part of Special Conditions —
                    Carpeting

No fill-in necessary.

Section 4.   Carpet Bid Specification Sheet
Indicate carpet color and number as selected by the company.

Type the company name, address, and the date in the appropriate spaces. Insert the correct page numbers on the Table of Contents page. Prepare a cover sheet.

Include a print of the completed finish-plan drawing for the facility, on which total square yardage and location of the carpet is indicated with the other bid documents, but do not bind it together with them. (See Construction Documents in the Master Forms Section for a sample plan.)

Bind all sheets together in the following sequence:

1. Cover Sheet
2. Table of Contents
3. Section A.
4. Section B.
5. Section 1.
6. Section 2.
7. Section 3.
8. Section 4.
9. Drawings

Distribute the complete bidding package to the selected bidders. Ask each bidder to pick up the package in person so that any questions can be reviewed at that time. Bids should be returned within a specific number of business days. When they are returned, review them and award the contract, unless the bids exceed the budget estimate by more than 5%. It they do, contact the officer in charge to determine why and discuss what to do about it.

When the Landlord Purchases Carpet

When the company's landlord purchases carpet as part of the lease agreement, most of the time he will have agreed to pay only up to a certain maximum dollars-per-square foot (or yard) cost. In such cases, the company will have to pay all costs over this maximum. Typically, a landlord's carpet allowance is less than the cost of a quality carpet. Therefore, the company may have a larger stake in the bidding process than does the landlord. As a consequence, be sure to be an active participant in the bidding process and retain the right to approve or reject any and all bids received by the landlord.

Since the company will not award the contract if the landlord handles the purchasing, it is not necessary to prepare a complete set of bidding documents. The company need only specify the type and color of the carpet selected and attempt to be involved in specifying the installation procedures. To do this, remove, copy, and complete the following pages from the Master Forms Section.

(Always keep in mind that the company does not want to undertake a responsibility properly belonging to the landlord. With that responsibility will come liability. Even though the company is paying a large portion of the carpeting cost, the landlord may object to your installation specification. Try to negotiate an agreement with him. This negotiation should be performed during the portion of lease negotiation involving the tenant work letter.)

Section 2.   Special Conditions
Fill in necessary information where indicated.

Section 3.   Installation Specification
No fill-in necessary.

Section 4.   Carpet Bid Specification Sheets
On this sheet, indicate carpet color and number.

Type the company's name, address, and the date in the appropriate sections. Insert the correct page numbers on the Table of Contents page. Note that Sections A and B should be deleted. Prepare a cover sheet.

Bind all sheets together in the following sequence:

1. Cover Sheet
2. Table of Contents
3. Section 2
4. Section 3
5. Section 4
6. Drawings

Include a print of the completed finish plan drawing of the new facility, on which total square yardage and location of the carpet may be estimated, with the other documents (but not bound together with them). Typically, the carpet installer prepares the estimate(s) of carpet requirements. The installer's estimates should be reviewed by the designer. (See Construction Documents Section for a sample plan.)

Assemble the complete package and submit it to the landlord. He will purchase the carpet and arrange for installation in accordance with the arrangements the company has negotiated with him.

Chapter 12

FURNITURE AND FURNISHINGS

Introduction

The commercial furnishings marketplace is full of various furniture lines in all qualities, looks and price ranges. It is a worldwide business with companies whose sales measure in the hundreds of millions of dollars per year. The U.S. manufacturers are the world leaders, both in terms of creative design and production. This gives the U.S. buyer an enormous variety from which to select in terms of aesthetics, functional design, quality, delivery times, availability of product and price. The market is staggering in size, variety and the complexity of the various selling and pricing methods.

It is also highly competitive and therefore highly volatile. Fortunately, there are a large number of individuals, dealers, general and specialty consultants, design and architectural firms who can provide guidance and current product knowledge. Use these resources wisely. Don't attribute all skills and all knowledge to any one group of individuals.

A periodic trip to the annual June NEOCON market in Chicago is a most worthwhile effort (and expense) for an active buyer. There are other markets in the U.S. and worldwide, especially Milan, Italy, for design.

Selection

The subject of selection is one the company will deal with as a matter of internal assessment. Remember that furniture serves not only a purely functional role, but along with the other design and construction elements, will affect the mood of employees and therefore their productivity, fatigue level, absenteeism and turnover. The company's selections should be comfortable, flexible, spacious enough for the equipment to be used and for the actual work to be performed, and as attractive as possible. It has been estimated that over a full life cycle of a building and its furnishings, over 90 percent of the costs are in personnel.

There are very few, if any, real measurement techniques for assessing work productivity as related to environment. There have been some studies performed, particularly by Steelcase, Xerox and others, on office productivity. Routine processing functions are the easiest to measure in terms of productivity. Not withstanding the lack of solid economic evidence in evaluating office productivity as it relates to office environment and the workstation, many designers, facility managers and other specialists in the office interiors industry are convinced that a correlation does exist. It appears that a pleasant, comfortable environment is more conducive to work productivity. Taken from the negative point of view, the fewer physical irritants or tiring conditions within the work environment, the more productive the worker is likely to be.

If, for example, by improving productivity or efficiency by a mere 10% (and the total labor force measured 100 people at an average of $35,000 per year), the total annual savings would be $350,000. The furniture is there only to serve employees. Employees *are* the company. Most likely they are the most costly expense to doing business. Give them the best designed and thought-out tools available. Once purchased, those tools must serve for a long time. Normally furniture requires little, if any, operating or maintenance expense. Employee payroll costs will increase every year.

Major Manufacturers and Dealers

There are five or six major commercial office furniture companies, and numerous comparatively smaller ones in the U.S. The major furniture manufacturers, such as Herman-Miller, Steelcase, Haworth, etc., are the industry leaders in product quality, reliability and often in service. They are usually the leaders in research and design in ergonomics, acoustics, workstation and seating usability and efficiency, comfort, adjustability and flexibility. Although most of them are not directly in the lighting or HVAC business, their research also involves these areas of the workplace.

As a buyer, the company is going to be faced with a very difficult series of decisions. The first is that often the major lines, both in size of manufacturer and quality of product, are strictly controlled. There may be only one dealer in the community authorized to handle one of the major lines. If that dealer is difficult to work with, too bad. You will either work through that dealer to acquire the line or not use the line. Sometimes a particular line will be handled through multiple dealers.

However, if the company will be a major customer, it may be assigned by the manufacturer to one dealer. The manufacturer will very likely assign a factory rep to work with you. If the company is a small or infrequent purchaser, it will probably have to rely on the manufacturer's dealer network.

The second problem with this traditional arrangement is that it tends to limit the price competitiveness between dealers for a particular line. In essence, what one pays for furniture from one of these sources is being controlled by the manufacturer through its selected dealer network.

There is one final problem to consider, particularly if the company is a large purchaser of furniture from one source. Even if you are not a national contract buyer, you may be considered a *house account*, assigned to a dealer for servicing. That dealer has little say in the pricing offered to you. You have no say in the dealer selection.

Once in a while, a non-authorized dealer may be able to buy the products you want from an authorized dealer. This is a practice called *bootlegging*. All of the major manufacturers oppose this practice, and if they become aware of it, may pull the line from their authorized dealer or stop the shipment.

Other Manufacturers and Dealers

The office furniture industry is one of the major industries in the world. There are numerous manufacturers, suppliers and importers of furniture marketing products which can range from superb to junk, and from very pricey to highly competitive. Very pricey products may not necessarily be good, and highly competitive products may not be junk.

Nearly all of these lines are sold through dealers, and usually there are a number of dealers from which to choose. This product variety may have its drawbacks. The comparatively smaller manufacturer may not be as financially stable as the larger ones. It is possible to order a sizeable installation, only to find that the manufacturer either cannot make the delivery on time or may even have sought bankruptcy or gone out of business. Before placing any large order, make some inquiries as to the financial condition of the manufacturer and its reputation for on-time delivery. If the installation is very critical, it may be wise to ask about the selected manufacturer's labor contracts and upcoming contract negotiations.

Dealers

The vast majority of the commercial office interiors industry functions through dealers. As in all businesses, there are any number of types and qualities of dealers. Some dealers may provide "free" design service, but they are not design firms. Some design firms are more decorating firms, and not actually dealers. These firms may purchase through an assigned dealer, who has added its markup and then sells the products to their customers at an additional markup.

Some dealers are so-called *stocking dealers*. That is, they carry a stock of furniture for the manufacturers they represent. Often carrying a stock of furniture is a condition to being a dealer for that manufacturer. That dealer must purchase the stock. It sits on the dealer's floor, usually as a non-producing investment, adding to the dealer's overhead, on the theory that it helps to sell the product. This theory is not without some merit, in that it may provide a place to see that dealer's lines in a convenient location and with salespeople who have product knowledge.

Other dealers are primarily *servicing dealers*. They rarely perform design, although they may take an active role in material and color selection. They are not usually stocking dealers. Their primary thrust is to provide service and pricing which reflects a lower overhead operation, using the buyer's design work, although they will assist in doing counting "takeoffs" of the actual products and number of pieces shown on the drawings. These dealers specialize in providing a maximum amount of service and problem solving before, during and after the installation. Rarely, if ever, will they represent one of the largest manufacturers, primarily because they do not stock.

The Architect's or Designer's Role

The architect or designer for the project is the most reliable and objective source of information on office furniture and furnishings. They are regularly involved in reviewing various manufacturers' current lines, but may also be aware of new furniture lines that may be available in time to meet the company's scheduled needs. Manufacturers expend large amounts of effort and money to keep architects and designers fully informed. Because of the architects' diverse and current knowledge, they should assist in making the primary and approved alternate selections. They will then write the specifications.

As already mentioned, some dealers are also design firms. There are occasionally some architectural firms who like to act as dealers. Almost any manufacturer will sell to an architectural firm and will do so at dealer's costs. This practice is not widespread, as most architectural firms consider it a conflict of interest as well as a function they are not properly equipped to provide.

National Contracts

Most major manufacturers offer a *national contract*. Whether this is an appropriate arrangement for the company will depend on a number of factors:

- How much furniture will the company be buying over the course of a year?
- How much of a discount will the contract provide? How will that compare to open market bidding, one manufacturer/dealer vs. another?

The use of national contracts tends to restrict you to purchasing only one manufacturer's lines. A national contract type of arrangement with a manufacturer, for systems furniture only, should not restrict you from purchasing furniture such as seating or files from any other manufacturer with whom the company may, or may not, have a national contract.

Many national contracts require the buyer to purchase a minimum dollar amount of a manufacturer's product during the course of the contract term (usually one year), in order to receive a specified, negotiated discount from the manufacturer's list price.

Some require that the buyer purchase products only from the manufacturer. The manufacturer will appoint a servicing dealer to install the products and pay that dealer a flat fee or a percentage of the contract amount for the installation. The manufacturer retains the relationship.

If your company is large and active enough, a national contract may be the ideal arrangement, especially if the company wishes to standardize on specific furniture lines throughout all of its facilities. This may be a preferable option due to in-house inventory and maintenance. A national contract can produce sizable savings from list prices and is somewhat easier to process than multi-product competitive bidding.

Delivery

Before placing the company's order, check with the dealer or factory representative (rep) for approximate delivery times, not only for the larger volume items, but also for smaller volume product. Those who are not very familiar with the commercial furniture industry may be surprised at the lead times for deliveries. Anticipate normal delivery times in the 10-15 week range. Coordinating deliveries, given multiple sources and different lead times, will be difficult. Most likely, the company will have to begin taking delivery of some products well in advance of others. The dealer usually has warehouse facilities where products can be temporarily stored until the company, as buyer, is ready to have them installed. If the company's storage needs are small and brief, this service could be free; usually it is not. If you are under a national contract, the factory rep will probably make arrangements with the installing dealer for this service, usually at your company's expense.

Shipping

Shipping or freight costs, as well as sales taxes, are additional costs most often paid by the buyer. Shipping can also add time to the actual delivery date—often a week or more. Ask about shipping costs and times. In addition to shipping or freight costs, there will very likely be cartage or delivery costs. The difference is that freight costs are over-the-road costs, from factory to a local warehouse, while cartage or delivery costs are from the warehouse to the final site. Often freight costs are borne by the factory as a *freight allowed* item and are a part of the factory's price to the dealer. Sometimes delivery costs are quoted as a part of installation. These are some of the things that are easily overlooked in budgeting and scheduling.

Quick Ship Programs

Most manufacturers maintain an inventory of finished products in limited styles and materials to provide immediate responses to moderate needs. These are called *quick ship programs*. Usually most manufacturers will be able to ship ten to fifteen workstations within one to two weeks. The selection of paint colors, fabrics and panel construction will be limited to the most popular, neutral offerings. Not all products will be available, but if time is critical, the service is usually available.

Pricing

For a company to determine how it wishes to purchase its furniture, it is important that it not only understand the marketing relationships, but also the pricing systems.

List

The first pricing term important to understanding the system is *list*. List price is the most current price for a particular piece of furniture or component quoted by the manufacturer to an authorized dealer. This price is always subject to change.

Some dealers are primarily design- or decorator-oriented. They may try to sell their clients on their design and decorating services and specify only furniture for which they are authorized dealers, or which they will buy from other authorized dealers. They will then sell the furniture and furnishings to their clients, often at full list price. If the designer is worth it to you and will make no other arrangement, at least be aware of the costs involved.

Deposits

Although not strictly a pricing matter, be aware that a sizable deposit may be required to secure an order. The amounts may vary, but will usually be in the neighborhood of one-third the cost of the order. This amount is often demanded by the manufacturer at the time the order is placed. The manufacturer is producing product to fill the company's specific order, not to inventory. He wants assurance that payment will be received upon completion of production and its ability to ship.

Discounts from List

In addition to the manufacturer's published list prices, it will publish a discount or discounts at which it will sell to a particular dealer. The dealer's profit comes from the difference between list (which the dealer will show the buyer) and the discounted price the dealer pays for the products. The amount of the discount passed along to the buyer will determine the cost to the buyer.

Most pricing systems are based on a 50% plus discount system. Many manufacturers quote their dealers a list price less 50%, less another 5%, 10% or even 20% depending on the size of the order and the pricing policy of the manufacturer. There is no real standard in the industry, although most manufacturers follow similar pricing strategies. A typical pricing practice is shown in Figure 12.1.

If the order is large enough, there may well be *factory pricing* provided in the form of additional discounts. Also, be aware that some very high volume dealers may receive an additional 1%, 2% or 3% discount. Governmental agencies will almost always buy at discounts which are greater than normal industry discounts.

One thing to keep in mind: no dealer is going to give you all of the discounts, as this would be giving away all of his percentage, which covers his own overhead and profit.

The various discount arrangements vary from manufacturer to manufacturer and sometimes from line to line within the same manufacturer. The important fact to be aware of is how the system works in order to acquire the best product for the best price.

Ordering Furniture Through a Bid Process

When purchasing furniture and furnishings made by manufacturers with whom the company does not have a national buying contract, order through a furniture contractor or dealer and bid the order. The first step will be to interview several prospective dealers, and select at least two of the best.

This is no simple task. There are a number of criteria in making these selections. Rarely will any single dealer be able to fulfill all of the desirable qualifications.

The following is a list of some of the more important considerations:

Reliability
Does this dealer have a reputation for getting the job done, even if it costs the dealer some of its profit or if it takes some extra or weekend time of some of its people? The company may need all of the help it can get.

Quantity	Items	Amount
1	Executive desk	$ 1,000.00
	Less 50%	500.00
	Net (cost per each $500)	$ 500.00
8	Executive desks	$ 8,000.00
	Less 50%	4,000.00
		4,000.00
	Less 10%	400.00
	Net (cost per each $450)	$ 3,600.00
80	Executive desks	$80,000.00
	Less 50%	40,000.00
		40,000.00
	Less 10%	4,000.00
		36,000.00
	Less 5%	1,800.00
	Net (cost per each $427)	$34,200.00

Figure 12.1

Accessibility

Does this dealer have all of the necessary service facilities within close proximity to the installation?

- Warehousing
- Delivery
- Installers and servicing personnel
- Repair and refinishing

Product Knowledge

Some dealers have vast product knowledge accumulated over years of working with many manufacturers and suppliers. Some dealers are primarily salespeople with marginal product knowledge. This is an area worthy of some in-depth conversations with the prospective dealers in order to establish their level of knowledge.

Pricing Philosophy

Many dealers establish one or more pricing formulas in order to secure the company as a long-term client. Ask if they will disclose their pricing formulas and work on a fixed markup over their direct costs to provide them with a reasonable margin to cover their overhead and profit.

The Bid Process

If the company intends to request bids for two or more competitive manufacturers' series, make sure you are bidding comparable lines. Not only should the lines be comparable in terms of quality, general appearance and design function, but they should have equivalent components. When doing comparative analysis, consider some non-economic issues such as the level of factory service available, the reputation of the installing dealer for service, and customer cooperation. Are manufacturer delivery times the same? Are all of the lines actually in production, or are there currently only prototypes? Are the lines being considered going to remain in the manufacturer's product line for a sufficient number of years to satisfy future infill needs?

Review and prepare the bidding documents by removing, copying, and completing the pages found in the Master Forms Section.

How much of the following document set is used will, to some degree, depend on the amount of furniture you will be purchasing and its cost. Regardless of how much documentation is utilized or how bids are developed, cover all of the points contained in these forms, in one way or another.

Section A.   Instructions to Bidders — Furniture Contractors

Fill in necessary information where indicated. Include two copies of the Proposal Form, one for the contractor to keep, one for the company to keep after the contract has been awarded.

Section 1.   General Conditions
No fill-in necessary.

Section 2.   Special Conditions — Furniture
Fill in necessary information where indicated.

Section 3.   Furniture Specifications
To specify the correct chairs, tables, free-standing desks, and the like, refer back to the final space plans. Using the information found there, complete Furniture Bid Specification Sheets and fill in all of the necessary data. Caution: this is very exacting work. What you specify is what you will get.

Insert the correct page numbers on the Table of Contents page, and prepare a cover sheet.

Prints of the completed Panel Plan and Furniture/Furnishings Plan for your facility, on which placement of items is indicated, must be included with the other documents, but not bound together with them. (See Construction Documents for sample plans.)

Bind all sheets together in the following sequence:

1. Cover Sheet
2. Table of Contents
3. Section A
4. Section 1
5. Section 2
6. Section 3
7. Drawings

Distribute the complete bidding package to selected bidders. Ask each bidder to pick up the package in person, so that any questions can be reviewed at that time. Bids should be returned within a specific number of business days. Be careful to allow enough time for the bidders to prepare their bids. If they do not have adequate time to review the bid documents and to contact their suppliers for pricing, their bids are very likely to be high, in order to cover unanticipated costs. When bids are returned, review them and award the contract, unless the bid exceeds the budget estimate by more than five percent. If it does, contact the person in charge of the facility to determine why and discuss what to do about it.

Ordering Furniture on a National Buying Contract

Find out whether the company has a national buying contract. If it does not, it may be efficient to attempt to negotiate one or more. To purchase furniture through this contract, remove, copy, and complete the pages found in the Master Forms Section and submit the package to the manufacturer's rep assigned to service the company's account.

Section B. Proposal Form and Instructions — National Contractors

Section 1.   General Conditions
(See Master Forms Section)
No fill-in necessary

Section 2.   Special Conditions
Fill in necessary information where indicated.

Section 3.   Furniture Specifications
Using the information found in your final Panel Plan and Furniture/Furnishings Plan and with the help of the manufacturer's rep, complete the Furniture Specification Sheets. To specify the correct panels, components, and other furniture, carefully cross-reference your furniture plans. Caution: this is very exacting work. What you specify is what you will get.

Complete the upper portion of each page. Insert correct page numbers on Table of Contents. Prepare a cover sheet.

Include a print of the completed Furniture Panel Plan and the Furniture Plan for the facility, on which placement of items is indicated, with the other documents, but not bound together with them. (See Construction Documents Section for a sample plan.)

Bind all sheets together in the following sequence:

1. Cover Sheet
2. Table of Contents
3. Section B
4. Section 1
5. Section 2
6. Section 3
7. Drawings

Chapter 13

ACCESSORIES, SIGNAGE AND GRAPHICS

Introduction

What are accessories? What are graphics? What is signage? Why should we be concerned about them, how do we use them and how should we acquire them?

Accessories are those small components which may (or may not) be a part of the company's overall philosophy of corporate image and function. They are things like ashtrays and desk blotters.

Graphics are interior logos, safety and life safety signs, identifying signage, directional signs, name plates and directories.

Signage as referred to in this book, is exterior signs. This is an arbitrary distinction and used solely to separate interior from exterior during discussions.

Each of these elements plays a significant part in providing identity, information and a sense of order to a facility.

Accessories

The most common accessory used to be the cigarette ashtray. In some offices and in certain sections of the country, it may still be a necessary accessory. It is likely that tip-action, wall-mounted or sand-urn-type ashtrays may still be required in lobbies and restrooms. The more common accessories today are such items as letter trays, PC keyboard trays and other computer accessories, leather or linoleum desk pads, wall and desk clocks, carafes and serving trays, calendar sets, rotary file units, memo pad trays, pen and pencil holders, and combination units providing a number of these items.

The more extensive the company's employee group, the more likely that the company will be buying accessories. As the need for uniformity of design diminishes, the closer accessories will come to office supplies. Office supplies are normally purchased for pure function, and their selection is based more upon price and availability. However, it may be necessary to pay some attention to accessorizing the offices and when doing so, pay special attention to those two twins of selection mentioned throughout office standardization – design and function.

There are four or five main manufacturers of accessories. Their pricing is usually competitive, and as a result you should be able to concentrate on design and function.

Accessories Selection

Working with the officer in charge and using the same selection criteria previously used in designating workstations, determine which of the company personnel are to get accessories and what their accessory package might consist of.

EXAMPLE:

Supervisors using a D-Station (and up through G-Stations) may all receive:

- Desk pad (not always required)
- Memo pad tray
- Desk calendar
- Paper tray and/or in/out boxes

E-Station personnel through F-Station personnel might receive all of the above items in addition to:

- Desk clock
- Additional paper tray
- Ashtrays for guests
- Water carafe and tray (not normally standard)

G-station personnel might receive, in addition to the all of the above:

- Pen and pencil holders
- Table lamps
- Coffee carafe, tray and cups (not normally standard)

Once this package has been agreed upon, it is suggested that a series of presentation boards showing pictures of the various items be put together. One board for each design line being considered, and separate boards for each personnel group. Using the same approval group used previously, try to establish a consensus for a design and an accessory item package for each of the personnel groups.

Do not be misled. Some of the employees will say something to this effect: "Oh, pick out anything. I'll be happy with whatever you select." They *do* care. You will hear about it if they do not have input in the selection of accessories.

How to Specify Accessories

When preparing to purchase accessories as single items or in small quantities, it is not necessary to prepare bidding documents. Simply contact several suppliers to compare prices, choose one and submit a purchase order, and follow the company's usual buying procedures.

Purchasing Accessories

The accessories supplier will probably have the items delivered directly to the company from the manufacturer. They will then have to be distributed to the appropriate personnel.

Be sure to:

a. Stipulate the date to be delivered;
b. Indicate the address and person to whom the order is to be delivered;
c. Include all quantities, manufacturer's numbers, finish numbers, colors, etc., as stated in the manufacturer's catalog and price sheets;
d. Complete the pricing information, including the unit price, the total costs of installation (if any), freight and all applicable taxes.

Graphics

Graphics is another of the corporate culture or philosophy questions which must be dealt with before the selection process can begin.

Graphics normally consist of such items as:

- Internal floor and department directories
- Directional graphics and arrows
- Safety markings on glass doors and partitions
- Department, division, and company signs within the office space
- Employee name plates on desks, workstation partitions, doors or full-height partitions
- Life Safety graphics and floor plans
- ADA requirements

To the extent possible, the above graphics should be coordinated with the building graphics, such as restroom door signs, electrical closet signs, etc. If the project is of sufficient size, consider replacing the building's standard signs with those of custom design. The cost will be small, and the design coordination will probably be worth the slight additional expenditure.

In the case of graphics the selection criteria is function first. Why do we need a sign here, what should it say, and where should it be placed? There are real values in good graphics, e.g., sensitivity to the company's visitors, a sense of order, planning and organization. Some signs may be required under ADA outlines. They may require specific text, fonts, sizes, contrasts, braille and specific mounting locations.

Graphics Design Help

The company's interior designer or architect is usually well qualified to help prepare the facility's graphics package. Often the larger architectural firms have specialists on staff who do nothing but graphics design. It may be a good idea to bring these people together with the company's own public relations personnel to develop a well-coordinated corporate identification program.

The graphics industry can supply the company with high quality design, use-consulting (where and how graphics should or must be employed), and specifying assistance. Once a supplier is chosen, ask for help. They will provide highly knowledgeable account and design people. Often this is a free service. Ask about it. Keep in mind that any custom design service provided by the supplier will probably be reflected in their prices.

Defining the Graphics Package

The first step in defining the company's needs is to determine how detailed a package is necessary to properly serve employees and guests. To do this, study the floor plans of the new offices. Are there multiple floors? If so, some directional graphics are needed.

Are the floors large and divided into departments? Again, some directional graphics will be needed, and possibly some departmental definition.

Will there be visitors such as salespeople, consultants, attorneys, accountants, and engineers visiting some departments?

Will there be visitors from related companies, from departments or divisions outside of the building visiting personnel or departments? Will there be a fairly high number of personnel who need intra-office mobility?

How much employee turnover is the company experiencing? How rapid are future expansion plans?

When studying the new floor plans and adjacency studies, try to envision how visitors and employees are going to find their way around. When thinking of visitors, plan not only how to get them from reception and elevator lobbies to the departments they wish to visit, but also how to direct them back to the elevator lobbies or the appropriate exit.

Formulating answers to the previous questions will help determine how sophisticated the company's graphics package needs to be.

To begin, secure a fairly clean drawing similar to 3-F2 or 3-F3 (see Construction Documents in the Master Forms Section.) Using a colored pencil, outline each department area as shown in the adjacency studies. Pay particular attention to the study showing the various departments and the circulation.

Begin with those departments or key people who receive visitors, either from the outside or internally, and which are the farthest away from the primary lobby. Work back to the primary lobby or exit. On the drawings:

a. Write down the name of the department or the individual.
b. In the circulation passageway at the entrance to the department, list the same department or key individual's name.
c. Move back toward the primary lobby or exit, to the next department, and list that department's name.
d. Repeat the first department name and the new department in the circulation passageway.
e. Continue this process all the way back to the primary lobby or exit.
f. At any merger with another circulation passage, make another list of all of the departments and key personnel accumulated to this point.
g. Repeat this process from all remote departments on the floor until you have listed all departments, both within their areas and in their respective passageways.
h. Prepare a final list of all departments and key personnel. This becomes the lobby directory.

Note: Do not forget the lobby itself. Help visitors to find their way.

A directory list will develop which will look something like Figure 13.1.

Once all the needed directional and identification signs have been defined, look around the actual space to see if there is any need for safety or warning graphics. Do not forget portable directories for special events or temporary signs required for maintenance, repairs or cleaning. Such items as floor-to-ceiling glass partitions are easily walked into unless identified by a glass drape or a graphics pattern placed to warn the unsuspecting. Pressure-sensitive decals can be used for this purpose. A nice touch is to have them done in the company logo or a pattern already present in the office environment. Identification of the stairways and elevator lobby are necessary, as are life safety floor plans in the elevator lobbies which show fire exits, alarm pull boxes, fire hose cabinets and a "You are here" map.

The final element is the lobby logo. Often this is overlooked or considered unnecessary. Whether this element should be included is part of the corporate culture or philosophy. Bring the question to the attention of the appropriate personnel and provide them with some idea of what sort of major element might be employed and what the costs might be. Again, the company's interior designer or graphics supply designer can provide ideas and estimated costs.

A good graphics package is not a terribly expensive component of the interiors package. It will add much more value than its cost in helping to tie together all of the planning of adjacencies, stacking and circulating on which so much time and effort was spent in the planning stages.

Signage (Exterior)

The distinction between signage and graphics is rather arbitrary in that signage is signage whether interior or exterior. However, in order to separate the two for the sake of discussion and description, this book refers to interior signs as *Graphics*, and exterior signs as *Signage*.

The primary difference between interior and exterior signage is the materials used in their construction and their lighting.

If the company is the major tenant or one of the major tenants in the building, it may want the company name and/or logo placed on the building or on the building grounds. If it is the only occupant in the building, there may be need for additional informational signage such as "No Parking" signs, "Stop" signs, "No Turn" signs, height signs in parking structures, "One Way" signs, exiting signs, and directional signs to entrance, truck docks, handicapped ramps and parking. Most often, these signs are the concern of the building architect. However, carefully review them for adequacy, keeping in mind the safety and security of employees and visitors. If they do not seem to do the necessary job, ask the landlord to supply additional signs or secure permission to have the company supply them.

At the Department Entrance	In the Passageway
PAYROLL DEPARTMENT	PAYROLL DEPARTMENT LOBBY
ACCOUNTING	PAYROLL DEPARTMENT ACCOUNTING LOBBY
AUDIT	PAYROLL DEPARTMENT ACCOUNTING AUDIT LOBBY

The next step is to add direction indicators to the passageway signs. The example might now look like the following:

→ PAYROLL DEPARTMENT ←	← PAYROLL DEPARTMENT LOBBY →
→ ACCOUNTING ←	← PAYROLL DEPARTMENT ← ACCOUNTING LOBBY →
→ AUDIT ←	← PAYROLL DEPARTMENT ← ACCOUNTING ← AUDIT LOBBY →

Figure 13.1

Signage Controls and Criteria

All of the considerations given to the lobby logo should go into the design and use of exterior signs. In addition to the design of exterior signs, keep in mind community sign ordinances. It is very probable there will be strict requirements as to the size, location and lighting of any exterior sign. The graphics designer or signage supplier may know this information. Otherwise, a call to the city planner's office or the city building department can usually provide the local regulations governing signage. It is also probable that the company's landlord will have signage criteria with which the company will have to comply. This information is often a part of the lease.

Chapter 14

PLANTS AND CONTAINERS

Introduction

Interior landscaping has progressed a long way from the days when someone brought in a prized African Violet and plunked it down on a desk. Interior plants are now taken for granted, although their acquisition, display and maintenance is hardly routine.

There are in excess of 150 varieties of plants which are now being propagated for interior landscape use. There is practically no habitable office space which cannot support live plants. (After all, mushrooms are grown in worked out coal mines.) Trends in architecture are providing larger floor plates, often with atriums that bring in large amounts of natural light. Artificial light from fluorescent lamps will provide all of the light needed for many plant species. Some species may be rotated into natural light during ordinary maintenance, or rotated back to the maintenance contractor's nursery for rehabilitation if they are normally placed in a below-standard light level. Finally, industry knowledge on what to specify — where, and how to maintain plants has become highly advanced.

There is no longer any reason not to utilize plants in an office environment, other than a corporate philosophy which dislikes plants or objects to their maintenance costs. There is no question that plants, whether leased and maintained or purchased and maintained, are an expense. However, even these costs can be substantially mitigated through the use of artificial plants and flowers. A large variety of extraordinarily lifelike artificial plants are available at quite reasonable costs and often can not be differentiated from live plants except by feeling the leaves or blossoms. Even these will require some amount of maintenance — cleaning, repairing broken branches, replacing missing leaves, and sometimes complete replacement from ultra-violet fading when placed in direct or reflected sunlight.

Increasingly, those of us who work in office environments are cut off from the outside. Often we rarely step outdoors except on weekends. We work in an artificial environment. We shop in an artificial environment. Our entertainment is likely to be at home, in a restaurant, or movie theater, etc. We have an increasingly more difficult time relating to the natural environment.

Plants, properly placed and specified, can add a natural element to our offices. They can and do serve as internal sound barriers and diffusers. They add a live texture to the often hard-line rigidity and monotonous repetition of many office layouts. They can offer life where there might otherwise be sterility. NASA's Dr. B. C. Wolverton has conducted extensive research on the value of live plants in our artificial environments. Dr. Wolverton's research shows that a large number of indoor air contaminants such as natural gas, gases from draperies, fabrics, carpets etc., can be reduced by placing live plants throughout the area. Even such chemicals as benzene, trichloroethylene (TCE), and formaldehyde are effectively reduced by live plants. Any number of commonly used products can contain one or more of these toxins: paper, facial tissues, ink, dyes, paints, plastics, rubber, grocery bags, cleaning agents, etc.

Fortunately some of the most common interior plants, such as Bamboo Palm, Mass Cane, Janet Craig, English Ivy, Spider Plant, and Gerbera Daisy are highly effective in reducing a number of noxious gases.

Most of us seem to feel better if we see a live (lifelike) plant now and again. As employers become more aware of the values of good work environments and their contribution to better work output, we will see more plant utilization, even if in limited areas.

Design

In as much as we cannot design buildings and office facilities around the use of plants, our designers should give some consideration to their use as the design matures from block diagrams to workstation arrangements. Plants can and have been used successfully and economically in lieu of partitions: as a dividing screen between departments; as screening around conference and seating areas and employee lounges; and as signals that there is a change in circulation.

Recently a number of design firms have begun to add qualified interior plantscape designers to their staffs. However, the majority of designs come from the plant supply firms. Given the present industry supply and maintenance system, this is not an all-bad source for design. More than likely, these designers have superior product knowledge and will be able to offer advice as to which species will do well in the available light; and what considerations should be given to the in-place HVAC diffusers. They also have a trained design ability to know which size and species of plants should be used to achieve the proper scale and massing in relationship to their settings. When it comes time to bid the company list of specified plants, do not permit a bidder to present a bid using down-sized plants (lower cost) to steal a bid. If you and your designers have selected a particular variety of plant in a 12" grow-pot, stick to a 12" grow-pot. Do not be talked down to a 10" grow-pot plant. You will not be getting a comparably sized plant or the same number of plants in the grow-pot.

Design Costs

Whether using an independent consultant designer or a supply company designer, the company will probably have to pay a design fee. The larger the job, the more likely that there will be a fee. Unless you can promise a supplier the job is theirs without bidding, the supplier cannot afford the cost of a designer's time on a speculative basis. We do not recommend buying large amounts of plants on a no-bid basis.

About Plants

There are some important things to know about plants and the plant supply and maintenance industry. Except for the truly independent interior plantscape designer or landscape architect, the industry is essentially two-tiered.

The first tier is the grower. Most American growers are centered in Florida, California and Texas, with Hawaii producing only a limited number of species for mainland use. The major growers are very large and independent of suppliers. They often specialize in certain species; however, most all of the growers grow the most popular plants.

The Growing Process

Most plants start as seedlings or cuttings and are grown outdoors, or in a growing shed in flats under a shade cloth and watered by an automatic irrigation system. When the seedling or cutting reaches a sufficiently stable size to survive in a little more hostile environment, it is transplanted to a grow-pot. A grow-pot is the black plastic pot plants are grown in at the nursery. Depending on the variety of plant, a grow-pot may contain anywhere from one to six or more plants. The grower will select the size of the grow-pot based on his estimate of future demand and on how large he wants the mature plant to be at the time he is ready to sell it. Rarely does the grower transplant the plant from smaller grow-pot to a larger grow-pot *and neither should you*. The odds of plants surviving a transplanting are less than 50-50. The plants you buy will be anywhere from one to six years old. They will live another two to six years depending on their size and variety.

The planting on view in the supplier's showroom was very likely grown in the same pot in which it started as a seedling. The size of the grow-pot determines the number of individual plants required to make a planting and the ultimate size of planting when available for sale and placement.

As the seedlings mature, the grower will provide nutrients, moisture and a shaded environment. Keep in mind that the tan you acquire in Florida is from sunlight in the range of 10,000 foot candles. It becomes obvious that plants must undergo some sort of acclimatization process before they can survive in the 100 foot candles of light near a desk. To do this, the grower begins a carefully programmed process of weaning the plant from the 10,000 foot candles of natural light down to an ideal survival light through the use of increasingly dense shade-cloths. These are graded in percentages; a 47% shade-cloth yields a 53% light. The grower knows how much sunlight is needed at any interval in the plant's growth cycle to be at optimal size for its grow-pot.

The Supplying Process

The second tier is the supplier. Today suppliers range in size from one-store, one-person proprietors to very large multifaceted operations sometimes owning their own growing operations. Most suppliers employ salaried buyers who buy directly from the growers for the direct account of the supplier.

At the time of sale, by the grower to the supplier, the plants may be given a spray of a topical coating to protect them from pests and ultraviolet rays during shipping to the supplier's greenhouse or storage facility. The supplier may clean and polish the plant of its grower-applied protective coating before delivering it, or may wait and clean it after placement.

Most suppliers have no way to tell if a plant has been properly acclimatized, and have no way of carrying out this process after receiving it from the grower. Suppliers are dependent on the reputation and skill of their growers.

Light Standards

As described previously, the grower must acclimatize the plant for its ultimate light environment. Today those standards are pretty well established. All of the plants available have been acclimatized, by variety, to one of the following light levels:

Low light plants	50 to 100 footcandles
Medium light plants	100 to 250 footcandles
High light plants	250 + footcandles

Plants of all sizes and appearances are available in any of the above levels. Some larger specimens may not be readily available and may require shipment from a grower.

Terminology

The following terms relating to plant size and shape are helpful to know.

Height:	Distance from the top of the soil in the grow-pot to top of plant.
Mean foliage width:	The widest part of the plant exclusive of any outstanding branches.
Foliage origin:	The point on the main stem where branching begins.
Cane height:	The height from the main stem to the branches or leaves.

How to Specify Plants and Containers

Plants

Plants should always be ordered through competitive bids. However, due to the variations in natural and artificial light, heating, cooling, heat transfer, drafts and humidity, it is important to determine which plant materials are most appropriate for each potential location in a facility. To do this, interview nurseries or interior plant suppliers in the area, and select three prospective bidders. Ask them to use their knowledge of local conditions, especially local natural light conditions and its effect on plants located near windows. Ask them to help prepare a list of appropriate plant materials and sizes, by grow-pot sizes, heights and mean foliage widths, which will be appropriate for the company's artificially lighted areas and those near natural light sources, such as windows or skylights.

Containers and Planters

Working with the company's interior designer and with the interior plant designers, select containers and planters for plants. It is literally vital to your plants that the appropriate size containers be selected for a particular grow-pot size. At the same time, an appropriate container must complement the plant and be consistent with the overall interior design and color scheme.

The container must protect the carpeting and furniture on which it is placed. Terra-cotta, a favorite with interior designers, is a major problem to those who must maintain the plants. Terra-cotta weeps. It sweats. It leaks. Plastic liners are available and sometimes fit. The interior plant designer and suppliers can show you a large selection of well-designed containers in plastic, fiberglass, metal, porcelain, glazed pottery, glass, even lined baskets.

The interior designer may have designed built-in or movable planters. These should come with well-made rustproof sheet metal liners.

Miscellaneous Materials

Once plants and containers are selected, the supplier will have to provide additional materials in order to properly install the plants into the containers or planters and apply the final top dressing. These sometimes constitute a hidden cost. Be sure that the bidders break these miscellaneous materials costs out so that you can know how much they will be.

- *Build-ups*: Foam blocks used to raise the grow-pot in the container to a proper height.
- *Liners*: Just as it says, a container liner to keep water from running out. Stainless steel, roofing paper, plastic.
- *Gravel*: Used to keep a plant from sitting in excess water.
- *Soil separators*: A separator between the top dressing and the soil in which the plant is growing.
- *Sterile fill*: Perlite or those little white foam peanuts used to pack dishes for shipping. They are used to pack between the grow-pot and the interior of the container to provide stabilization.
- *Top dressing*: Wood bark, wood chips, or peat moss used to provide the final dressing on top of the soil separator.
- *Plastic liners and dishes*: Plastic dishes usually placed under the container to catch excess water and condensation.

Maintenance

Normally the plant supplier proposes to provide a maintenance contract for at least the first year. This is important as it is the means by which the product warranty will remain valid.

Under a typical Maintenance/Warranty Contract, you should expect the terms shown in Figure 14.1.

What *you should not do* if you have a Maintenance Contract is outlined in Figure 14.2.

The Bidding Process

Using the best suggestions received from the three prospective nurseries or interior plant suppliers, develop one list that meets the approval of the coordinator and the interior designer. Make this single list the basis for the bids by inserting it in the Plants and Containers Specifications Sheets found in the Master Forms Section.

Typical Maintenance Contract Terms

a. Regular and sustaining watering of plants.
b. Spraying and controlling of insects and diseases.
c. Cleaning and polishing plant foliage.
d. Rotation of plants to lighting sources.
e. Trimming plants.
f. Feeding and monitoring plant nutrition.
g. Replacing unhealthy plants with a plant of the same size and variety or a plant of equal value and size. This may not occur if the original plant is unavailable or deemed not appropriate for its location. Remember, most plants will have only a three-to-six year life. Part of the maintenance agreement is, in essence, a replacement contract.

Figure 14.1

Next, review and prepare the balance of the bidding documents by removing, copying, and completing the pages in the Master Forms Section as follows:

How much of the document set is needed will, to some degree, depend on the amount and cost of the plants and containers being purchased. Regardless of how much documentation is prepared or how the bids are developed, cover all of the points contained in these forms in one way or another. Do not trust the contractors' reputations. When it comes to a loss, the company's money or his, have no doubt about whose money the contractor is going to protect.

Section A.   Instructions to Bidders
Fill in necessary information where indicated.

Section B.   Proposal Form and Instructions
Fill in necessary information where indicated. Include two copies of this form, one for the nursery to keep and one for the company to keep after the contract has been awarded.

Section 1.   General Conditions
No fill-in necessary.

Section 2.   Special Conditions
Fill in necessary information where indicated.

Section 3.   Plants and Containers Specification Sheets
Fill out these forms using the list of appropriate plant materials and quantities. Type the company's name and address, and the date in the appropriate places.

Insert the correct page numbers on the Table of Contents page. Prepare a cover sheet.

Include a print of the completed Plant Location Plan for the facility, on which placement of plants is indicated, with the other documents. (Do not bind it together with them.)

Practices to Avoid Under a Maintenance Contract Agreement

a. Water plants.
b. Treat plants for insects, pests or disease.
c. Fertilize plants.
d. Lower light levels below those originally specified or reduce the number of hours of light.
e. Trim or prune plants.
f. Introduce any liquids into the plant soil such as stale coffee, alcohol, sugar, cleaning chemicals, etc.
g. Permit toxic gases in the area of the plants such as strong ammonia or floor-stripping chemicals.
h. Permit ambient temperatures to exceed 85°F or fall below 55°F.
i. Relocate, move or rotate plants.

The company is expected to provide or permit:

a. Access to hot and cold water, preferably in a janitor's closet.
b. Access to plants during normal working hours.

Figure 14.2

Bind all sheets together in the following sequence:

1. Cover sheet
2. Table of Contents
3. Section A
4. Section B
5. Section 1
6. Section 2
7. Section 3
8. Drawings

Distribute the complete bidding package to the three selected suppliers. Ask each bidder to pick the package up in person, so that any questions can be reviewed at that time. Bids should be returned within a specified number of business days. When they are returned, review them and award the contract, unless the bid exceeds the budget estimate by more than 5 percent. If it does, contact the coordinator immediately to determine why, and to immediately discuss the problem, and what to do about it. A number of solutions can be reviewed. Working with the designer and the most likely supplier, consider eliminating certain plant locations. Consider reducing pot sizes and therefore plant sizes, or changing to alternate plant species which are less costly. Whichever solution(s) you choose, plants are worth the effort.

Chapter 15

ART

Introduction

Defining a company art program is often like trying to grab a handful of smoke. When it is there, you can see it. Often it has a discernible taste and can produce an emotional response. However, trying to define a company art program, good or bad, is a very difficult undertaking.

Art has been around since the dawn of mankind, and although it is essentially a means of communication, it requires a special form of skill and execution. However, art exceeds mere communication by adding the element of beauty or emotion. Two individuals may never agree on a single definition of beauty, nor on what sort of emotion art should evoke; but they probably would agree that without beauty or emotion, it is not art to us. A lot of other qualifications could be added to this simple definition, but for our purposes, this definition will suffice to establish a basis for a company art program.

Out of this definition come questions which will be key elements in establishing an art program. What is being communicated? What is the language? Who can understand it, and does it suit its audience? What is the beauty? Again, from the standpoint of our audience, is the beauty perceptible? Is it acceptable? We have two more questions to consider: who is our audience; and what is the company role, if any, in presenting art?

Why Company Art?

If you are the coordinator or officer responsible for the project, give some serious thought to the question of what kind of company art program there should be. How should it relate to company culture and philosophy – public and internal? What is the company culture and image philosophy, public and internal? What form should the art program take? How can it be implemented? What value does it have for the company?

First, recognize that employees and other viewers come from highly diverse backgrounds. Those from one set of socioeconomic experiences may be attuned to certain ideas and points of view reflecting their backgrounds. Other employees and viewers from different socioeconomic experiences may have entirely different cultural exposures and may have completely different interpretations of art and beauty. Each of the many groups and individuals comprising the company's employees and viewers will experience unique responses to the art program. In fact, they will probably have very different responses to colors. Even the geographic location of the office will have a strong cultural influence on the tastes of most all of the viewing audience.

Second, communicate with senior management in an attempt to offer views on the roles and attitudes of a company art program – and to learn senior management's views. Without the interested support of senior management, there will be no significant art program.

There are three employee-related values which can result from an art program.

1. *Stretch the Mind.*

   There is little question that a company art program that extends to some degree into most work areas will have direct effects on employees. It will tell them that management cares about them as persons as well as workers. As the art program expresses an image, it will help them to identify with the company's image of itself. For example, it will communicate to the employees that this company may care about the society in which it functions, as well as being an institution to make money. Employees who have higher self images will have better work attitudes and therefore better work productivity.

2. *Educate.*

   A dynamic art program should have within it some mechanisms directed at educating the viewer, employees or visitors about the selected pieces and their artists. Some corporations carry these efforts into actual formalized lectures and publications on "... about this piece or why this is a good piece;" or "...what is this artist saying to us," often extending beyond the company's art works into a history of the various types and schools of art. Programs of this nature yield improved acceptance of new ideas amongst employees. There is an extension of the employee's thinking horizon, with a perception of self-improvement and self-image which will result in a greater loyalty. "I am more than a number to this company, I am an important person with recognized intellectual capacities."

3. *Create A Healthier Work Environment.*

   A well-structured art program should act as a subtle relief from the high stress work environment and provide a few moments of diversion. A momentary view of tranquility, or perhaps an alternate form of stimulation, will be an intellectual stimulant or relief. Most psychologists confirm that the average person functions at high levels of concentration for about forty minutes at a time. After forty minutes, the brain looks for a few minutes of diversion. Some employees daydream; some begin to chat with their immediate neighbor; some call home; others doodle. Some stare out of the window or at a piece of art. Take your choice.

Level of Art Commitment

As the company (and its consultant) work through the selection process, it will find itself in one of four levels of commitment. Management will ultimately decide which level it is in and into which level it proposes to fit. These classifications may not be the most appropriate for all companies, but they can be a guide.

Blind

What do we need art for? We are successful in business. We have never had an art program before. We will continue to make money and be successful without art. Why art?

Color on White Walls

All we need art for is to fill up some blank walls in our public spaces. Maybe a couple of pieces in the boardroom, the conference rooms and some of the executives offices.

Middle Ground

If we are going to do this, let's do it right. Let's use some local artists' works as well as non-local artists' works and spread it throughout our offices as well as in the public spaces. But, let's keep within the budget.

Investment Grade

As long as we have a program, we should build something of cultural and economic value for the future.

It is interesting to note that although the company may fall into one category or another, once a program has been started, a company usually progresses. Rarely does it regress.

The Art Consultant

If management has decided to "...buy some art," the coordinator must now find a qualified company art consultant. Finding a good art consultant is a must. This is no field for the do-it-yourselfer, or the employee with a couple years of art history in their educational background. The good company art consultant is tough to find. There are no associations that we know of. There are no degrees being offered directly in the field. There are no certifications and no licensing. There are, however, some criteria which are important for reviewing the consultant's qualifications. The following are some of the available sources and their relative strengths and weaknesses.

Commercial Interiors Dealer

If the purpose is to merely fill a blank wall in the reception/waiting room, the company's commercial interiors dealer may be willing to satisfy this need for you. The representative may not be very knowledgeable about art but will likely have a good sense of color and scale and should know the company's tastes. Their compensation will come as a percentage markup on anything they buy for the company.

Framer/Gallery

The streets are full of framing shops, many of which also sell a variety of prints and posters. They may have a friend or a staff person whom they wish to offer as a consultant. This could be a beginning. It may be worth talking to the suggested consultant.

The framer/gallery can offer a vast variety of prints from a number of major publishers of artists' prints and posters. If prints are the extent of the company's art program, the price of the prints and their framing will be the major criteria for selection. The gallery operator may be able to speak "art," but be very careful in assessing this person as an art consultant.

Gallery/Dealer

Less frequent, but still to be found on the street, are the gallery/dealers. These operators usually have a group of artists whom they represent. They will have exclusive or nonexclusive representation rights for the works of particular artists. The better known the artist, the more likely the dealer will be to have exclusive sales rights for a particular area. Today many better known artists no longer sell their original works, but rather sell only prints or serigraphs, which are a form of silk screening, but personally produced by the artist. The gallery/dealer may have many acquaintances in the art community and could possibly direct you to an independent consultant. The gallery/dealer will have a certain stable of artists whose works it will wish to sell. These artists and their work may provide the basis for the beginning of a collection; however, they will, by necessity, be limited in total art spectrum and may or may not be suitable for the company's needs and desires. This group is compensated by virtue of their sales of art works of artists they represent.

Dealer/Artist Representative

The dealer/artist representative is similar to the gallery/dealer, but without a display gallery. The artist's representative will have a strong bias toward his stable of artists and, like the gallery/dealer, very often will have a predisposition toward a particular genre of art. They too will be compensated by receiving a commission from the sale of works by artists they represent. If the particular artist or genre of art is one which will fit the company's present art program, work directly with the dealer.

Independent Consultant

The most desirable, yet most difficult to find, is the qualified independent company art consultant. There seem to be two reasons for this: the limited market for their services often drives them into one of the other categories described above, and concern with their capabilities to deliver a challenging, well-conceived, well-communicated program within budget, which will meet management's art concepts. The woods are fairly full of individuals who can deliver parts of a good program, but are lacking in other parts. In other words, many consultants know art, but are not totally qualified to undertake and execute a professional, unbiased, structured program which is sensitive to the company's offices and image. When you find one, they are worth their weight in gold!

Some criteria to use in interviewing and assessing prospective consultants are:

Background in Art

Ask for a vitae or resume. Whether a vitae is available or not may be a clue as to the candidate's level of professionalism. Formal training in art history, or as an artist, can provide some basis for evaluation. How fixed the candidates' opinions are on schools or genres of art may indicate whether a recommended program will best suit the company's ideas and needs, or is simply of special interest to the consultant. The more comprehensive the candidates' backgrounds, the fewer biases may be presented to the company.

Professionalism

Since this is not a very well organized or defined profession, professionalism is a qualification which will be hard to define. How well organized is the candidate's presentation? How well planned out is the process by which this candidate proposes to proceed? We suggest the following plan:

1. Establish a clear understanding of the client.
2. Establish a clear understanding of the client's specifications, level of investment and art concepts.
3. Interview client user departments.
4. Produce a written Art Program.
5. Conduct research.
6. Determine appropriate images.
7. Review Cost Feasibility (budget), with the client; if approved, proceed to step 8. If problems, return to step 5.
8. Present visual and written proposal conforming to program; if approved, proceed to step 9; if problems, return to step 5.
9. Secure client approval.
10. Acquire art.
11. Frame art.
12. Deliver art.
13. Install art.
14. Art data documentation.
15. Monitor program

Process

Whether the candidate has structured a program similar to this or not, the company should have one. Can your candidate fit into a structured program such as the one just outlined, or is the candidate likely to want to talk a little and then run out and buy a lot?

Communication Skills

Notice that the first three steps outlined concern the consultant's ability to listen and to articulate not only what the client says it wants, where it stands, and what it perceives as its art image, but also to elaborate on and complement those thoughts.

Verbalizing the Program

Does the candidate propose to provide a written program defining what the company's objectives are; what the consultant is going to do; and how the job is to be done?

Budgeting

Does the consultant candidate agree that budget parameters can be adhered to? If there is not a fixed budget, how does your candidate intend to aid in establishing one?

Documentation

Can the candidate offer a sample of the sort of art data documentation which is to be provided with the package? Preparing a catalogue of the art works is a lot of work, particularly if done properly. It should be possible to judge the level of professionalism of the candidates by seeing a sample of an existing catalogue. Do not be surprised if the candidate cannot produce this document. If not, and you are otherwise favorably impressed, ask the candidate to produce a sample catalogue.

Follow-up and Monitoring

What sort of follow-up does the candidate propose? Look for offers to personally explain pieces to the employees in whose spaces the art is placed. Look for suggestions to relocate certain pieces for variety, or to follow some change in personnel or function.

Compensation

We have outlined above how all of the non-independent consultants would generate their compensation. The independent consultants will most likely operate on any one of the following basis:

   a. A negotiated fixed-rate contract. This may be based on an estimate of time to be spent on the project or a percentage of the art value, or on what the consultant thinks the traffic might bear.
   b. A percentage of the procured art, framing and hanging costs.
   c. The consultant may arrange for a percentage of the artist's sales price or the artist's agent's commission.
   d. An hourly rate with a fixed maximum. This is the author's preference because it yields the highest degree of independence from market and personality influences.

Whatever formula is decided upon, be sure that it is carefully defined in the contract. However, be fair. If some of key company people are consuming inordinate amounts of the consultant's time, either intervene to protect the consultant or arrange for additional compensation. The subject of art is a wonderful topic of conversation.

Company Art Concepts

Company Image

At a number of points in the total design process, the question of company image has been mentioned. Often this image, if not thought out and defined, will be the mere outgrowth of an interior designer's personal tastes or the designer's execution of whatever is in vogue at the time. There is nothing fundamentally wrong with being in vogue as long as the particular style is in keeping with the nature of how the company views itself and how it would like to view itself.

How does the company believe it is viewed by its customers or clients, or by its peers; by the community in which it resides, or by its employees? These questions should be carefully considered during the entire design process and will be most graphically expressed in the final art program.

One of the great things about an art program is that a good consultant can blend seemingly incongruous styles to result in more than one image, often with dramatic results. For example, an image of high-tech, success and strength might blend nicely with art styles which are outgrowths of the impressionists. A very traditional image might support carefully selected and displayed abstractions.

Whatever image is decided upon and whatever art styles are selected, they should represent some sort of consciously thought-out statement. It should consider all of those many elements of the community which will be viewing it, and in consequence, viewing the company itself.

Budgeting

Earlier in this book we dealt with the preliminary budget and final budget. Both of these budgets should be completed and have appropriate amounts allocated for art before their final approval. There are about three approaches to establishing a budget for art work.

Percentage of Project

One of the more commonly used formulas is a percentage of the total project budget. This can range from one to two percent, depending on how much art will be relocated, if any, and which level of commitment management decides upon.

Dollars per Square Foot

A range of $1.00/SF to $4.00/SF is also a reasonable criteron, and seems to be within the reported national statistical range, but only as long as the project is in the moderate size range. If the project is extremely large, you may wish to moderate this approach.

Actual Needs Estimate

The third approach involves bringing the consultant on board during the preliminary budget process and walking the job through on paper from the design development drawings. The consultant will review all of the walls in the office space and decide which walls will receive art and which will not. An arbitrary value, which matches the commitment level, is then assigned to each wall. The amounts are totaled, and amounts added for framing and hanging.

The consultant's fee should not be included as a cost of the art but should be budgeted along with the other consultants, such as interior designers, architects, lighting engineers, and the like.

Scheduling and Installation Process

Scheduling

Once the budget has been set and space planning has begun, the art selection and acquisition schedule is established – in reverse. The target date for move-in sets the date for installation. The actual *art-in-place* date should be only a few days prior to the open house date, if scheduled. If there is to be no official reception date, the art-in-place date should be a few days after the move-in has been concluded and most of the chaos has settled down. Fine art pieces should not be subject to the rough activities of the actual move-in.

The consultant will establish a schedule, working backwards from the target date. The second advantage in this procedure is the opportunity for the consultant to have better coordination with the design team and their finalization of space utilization, preparation of the finish plans, color selection, and lighting. The art consultant must be on-board and active early enough to respond to design activities and to influence design, color and lighting when necessary. The consultant should also begin the walk-through process, reviewing actual wall surfaces for light, color and viewing range. It may be logical to place a large piece of art on a large wall, but not so logical if the viewer can only get a few feet away from it in order to view it.

Interviews

To begin the actual selection process, there should be some interaction between the consultant and key supervisory personnel in areas where works will be placed. The consultant should understand the activities and the types of employees or visitors within each area in order to select appropriate works.

If executives are to have any input into works selected for their offices, they too should be interviewed. This offers an opportunity for the consultant to explain to the individual executives the nature and direction of the program, so that executive selections will fall within the program.

Research

Once the consultant has been selected, it is necessary for a key person to be selected by the company to act as the liaison between the consultant and other company personnel, and to be able to approve or disapprove of the many decisions which the consultant will be trying to make on the company's behalf. The key person should be supported by a small, involved committee. This committee should represent diverse levels within the company. If the art is to be spread throughout many areas, the committee's input will be challenging and assist both the *Key Person* and the consultant in helping to define the objectives of your corporate "art." The key person need not have an art, color or design background, but should have strong leadership abilities and have displayed a creative, open and inquiring mind.

Upon completion of the interview process, the consultant should review the findings with the key person, pointing out any potential difficulties or conflicts. Once these questions are clarified, the consultant begins the research phase, resulting in the visual and written presentation. The same sort of review with the key person takes place. If necessary, a re-submission takes place, and a final budget is approved.

Approval

Authorization for acquisition should be granted to the consultant. If commissioned works are involved, a pro forma (advance) payment may be required.

As works are received, they should *always* be acquired on a Payment on Approval basis. Do not be hesitant about this procedure. The company does not wish to acquire art that is not what it had understood it to be. This problem is a bit more difficult when acquiring commissioned art. However, if the work substantially departs from the artist's sketches, and it is not going to fit the program – reject it. The probabilities are that a good consultant will have already dealt with this problem, on the company's behalf, in which case it will not be an issue.

Installation

Installation has already been covered in the review of scheduling. It is more important to have the work in place at the appropriate target date, than to become overly concerned about damage during the fine-tuning stage after move-in.

The installer should provide anti-theft hanging systems as a matter of course. Ask about them; do not assume that they will be there.

Documents

Data Documentation

Fine art works should each come with some form of certificate of authenticity from either the dealer or the artist from whom they were acquired. This document may consist of a Bill of Sale, a Fine Arts Appraisal, a formal Certificate of Authenticity, a Warrant and Guarantee, and a Letter of Description. There is no set document form that is universally followed. Whatever documents you receive should describe:

- The name and address of the dealer (could be a letterhead).
- A brief two- or three-word description of the work.
- The media in which it is executed (sometimes).
- Size (sometimes).
- Its title (if any).
- The artist's name.
- The date of execution (sometimes).
- The date acquired.
- Purchase price (this is often a separate document).
- A statement that the work is "authentic" or "original".
- Signature of the dealer or artist.
- Date of the document.

If the work is of major importance or is an old piece, documentation should also include a warranty of ownership and possibly a provenance or abstract of prior ownership.

Finally, there should be information about the piece of work, if it has any historic significance; and a biographical statement about the artist.

The consultant should assemble all of the material and may have a standardized form including much of the above information.

Catalogue

Based upon the documentation, research and knowledge, the consultant should prepare a catalogue of the company art. It should contain all of the above data obtainable, together with a clear photograph of each piece of art. In addition, the consultant or someone in the company may wish to prepare a brochure describing the various works, their histories, what the pieces represent and some backgrounds on the artists. This brochure can be a part of the company's "Welcome to Our New Offices" memo.

A brochure describing the company art program and the specific works of art can be a good public relations piece at an open house or handed out to the media. It will serve to point out the company's interest in art, employee relations and the community. Have in mind that the preparation of a brochure of this sort is not within the normal scope of work for the consultant and there may be additional compensation costs for it. Nonetheless, the consultant can probably produce this work at less cost and in less time than the company could do it in-house.

Monitoring and Ongoing

Once established, art programs often take on lives of their own, particularly if documentation has resulted in a program that is active and relates to the community.

After the works are in place, do not be afraid to consider relocating or rotating various pieces. This will provide a fresh look in some areas and permit employees in different areas of the offices to view a variety of pieces. As with most museums, the company need not display all its art at any one time.

Art in the company's offices is there to perform a function; let it work for the employees and for the company.

Chapter 16

LIGHTING

Introduction

Office building design has evolved over the years to reflect the more specialized demands of the office space user. A few years ago, it was sufficient for our space planners to design a lighting solution consisting of rows of building standard fluorescent 2' X 4' fixtures on 8' or 10' centers which yielded about 100 to 150 footcandles of sustained lighting. Since the energy crisis of the mid 1970's, the more recent proliferation of computer terminals and a better understanding of the relationship between office working conditions and productivity, office lighting has become increasingly critical.

As lighting requirements have become more specific, the lighting industry has been responding with more specific solutions. Lighting engineers, in particular the Illuminating Engineering Society (I.E.S.), have developed more scientific methods of articulating the need for the quality as well as quantity of light. Today, lighting of office space is no longer one of brute force, i.e., "100 f.c. everywhere," but has become so complex that even the lighting engineers must use computers to produce appropriate solutions to the variety of tasks to be performed and the space conditions.

The purpose of this chapter is to provide a basic understanding of light, task-related considerations, and the challenges inherent in lighting.

Basic Lighting Requirements

Given the currently available technical solutions to lighting, we should begin with some basic industry-accepted standards, such as the IES Illuminance Categories (shown in Figure 16.1).

Figure 16.2 lists selected Illuminance Categories, from the foregoing chart, which relate to office space areas and activities. The illuminance categories indicate rather broad ranges. These ranges anticipate variables in task importance and average age of workers. There is strong evidence that the light gathering capabilities of the eye drop off quite rapidly, beginning at about 40 years of age. Corrected vision only partially adjusts for this loss. As the average age of workers gets older, the amount of light required for them to perform with speed and accuracy must increase.

Consultants It is a good idea to engage a qualified lighting designer or engineer (IES) to design the company lighting systems. When considering lighting designers or engineers, be sure to inquire as to their recent activities in ambient lighting. Likewise, budget for a quality system that is flexible and responsive to various problems, tasks and spaces; and is interactive with schedules, load shedding, and solar influences – in other words, a system that manages the company's lighting. The premium cost for such a system should be in the $1.00 to $1.50 per square foot range. The energy cost savings should repay the premium in 1-1/2 to 3 years. The labor productivity improvement could be substantial.

Basic Lighting Standards		
Activities	**Illuminance Categories**	**Range of Illuminances (in footcandles)**
General Lighting		
Public areas, dark surroundings	A	2-3-5
Simple orientation, short visits	B	5-7.5-10
Spaces where visual tasks are occasionally performed	C	10-15-20
Illuminance on Task (Typical Office Tasks)		
Visual tasks of high contrast or large size	D	20-30-50
Visual tasks of medium contrast, small size	E	50-75-100
Visual tasks of low contrast or very small size	F	100-150-200
Illuminance on Task Combining General and Supplemental Lighting (very unusual tasks)		
Visual tasks of low contrast and/ or very small size for prolonged periods	G	200-300-500
Very prolonged performances exacting visual tasks	H	500-750-1000
Very special visual tasks of extremely low contrast and small size	I	1000-1500-2000

Source: IES Lighting Handbook, 1981

Figure 16.1

Illuminance Categories	
Area/Activity **Illuminance**	**Category**
Conference Rooms	C
General Office Spaces	D/F
Private Offices	D/F
Lobbies, Lounges, Reception Areas	C
Reading Areas of Libraries	D
Microfiche	B
Photographs with moderate detail	E
Xerograph	D
Data Processing Tasks	
VDT Screens	B
Impact printing (good ribbon)	D
Keyboard reading	D
Data Processing Rooms (active)	D
Educational Facilities	
Classrooms	D
Laboratories	E
Lecture rooms	F
Chalkboards	E
Felt-tipped pen – marker boards	D
Food Service Areas	
Kitchen	E
Dining	B
Cleaning	C
Service Spaces	
Corridors and Stairways	C
Elevators	C
Washrooms and Toilets	C

Source: IES Lighting Handbook, 1981

Figure 16.2

Definitions

To deal with some of the following material, it will be necessary to establish some definitions. (See Figure 16.3.) It may not be necessary to learn how to execute the formulas used in these definitions, but it helps to be aware of what they involve. Most interior design consultants know very little about lighting. It may be necessary to deal with the lighting designer or engineer directly and in concert with the company's architect or designer. These definitions have been simplified as much as possible, and while they will probably be sufficiently complete for the project manager's needs, they may not completely satisfy a lighting professional.

Lighting Definitions

Ambient: The surrounding environment. Ambient lighting is usually a universal lighting system that illuminates large areas to a universal level of light.

Candlepower (CP): Literally, the illuminating intensity of a standard candle. The intensity of light produced by a light source, usually measured in a specific direction.

Cavity: In this context, the cavity of a particular room or a particular portion of a room, usually from the work plane to the ceiling.

Cavity Ratio (CR): An index or proportion involving a calculation of a room cavity: (usually measured from the top of the work surface to the plane of the lighting source).

Color Rendering Index (CRI): An International Commission on Illumination scale for measuring the ability of a light source to render accurate color. The scale is from 0 to 100, with 100 being the most accurate representation.

Coefficient of Utilization (CU): The amount of light reaching the work plane, described as a percentage.

Diffuser: Usually refers to any light shielding part of a fixture through which the light passes. Typically, a plastic or glass lens or prismatic panel with geometric shapes molded into it to direct the light down. A true diffuser is a milk-white panel that has a uniform brightness.

Efficacy: The number of lumens of lamp output per watt of electric power.

Equivalent Sphere of Illumination (ESI): A "seeability" measurement. Sphere illumination is the amount of light, in footcandles, which falls on a task from all directions, as if the task were placed in the center of a sphere of light. The ESI is the amount of light which needs to be emitted to provide an equivalent task visibility, "seeability", in a situation. If one wished to achieve a 75 ESI f.c., one might need to generate from 70 f.c. of excellent quality light, (with no veiling refections) to 300 f.c. of poor quality light (with severe veiling refections). This is usually a computer calculation made by the illuminating engineer.

Fixture: The device built to hold, operate, project and protect the light generating lamps. A luminaire.

Footcandle (F.C.): The amount of light generated by one standard candle at one foot distance. A measurement of light. One lumen projected on a square foot equals one footcandle. Footcandles = lumens/square feet.

Footlambert (FL): A measure of brightness, direct or reflected, as would be seen by the human eye. In general, one lumen leaving a source area of one square foot would equal one footlambert, in the direction of view.

Glare: Extreme brightness whether direct or indirect.

Indirect Lighting: Lighting from sources which cannot be directly seen, such as from a source aimed at the ceiling from suspended luminaires.

High Intensity Discharge (HID): A group of lamps which include Mercury-Vapor, Metal Halide and High Pressure Sodium.

Lamp Lumen Depreciation (LLD): A percentage averaging factor used to account for the gradual decline in lumen output of a lamp.

Lens: See Diffuser.

Figure 16.3

Louver: Another type of shielding device for a lighting fixture. A plastic or metal baffle with geometric shapes formed into it in such a way as to direct the light downward from 0 degrees vertical to 30 or 45 degrees from vertical.

Lumen: A measure of luminous flux emitted from a source.

Luminaire: Another name for a light fixture.

Luminaire Dirt Depreciation (LDD): A percentage factor used to adjust lighting output for an accumulation of dirt and dust. Depends on the fixture type, general space cleanliness, and especially air-handling filters. The factors typically range from .75 to .90, much lower for high dirt or dust areas such as factories or outside lighting.

Luminaire Efficiency: The number of lumens actually emitted by a fixture divided by the number of lumens generated by the lamps in the fixture.

Maintenance Factor (MF): A combination of luminaire dirt depreciation (LDD) and lamp lumen depreciation (LLD).

Parabolic: The geometric shape used in many louvers to direct light to the geometric axis of a parabola, with a total result of redirecting light from angles that produce glare.

Plenum: A space usually between the ceiling and the structural deck above. This space usually contains the lighting fixtures, lighting wiring, air-handling ducts, sprinkler lines, telephone and data lines. It often acts as the return air chamber for the air circulation system.

Polychlorinated Biphenyl (PCB): A highly carcinogenic substance used in magnetic ballasts and transformers, as dielectric, mixed with tars to isolate and sound dampen the internal plates.

Reflectance: A percentage measurement of the ability of a surface to reflect light. An acoustical ceiling might have a reflectance of 75% to 85%. Light grey painted walls would have a 45% to 75% reflectance. Most commercial interior paints have a reflectance factor indicated on the paint color chip.

Specular: A mirror surface in a lighting fixture, usually metal (although some fixtures use a mylar specular film instead of polished metal).

Troffer: A lighting fixture recessed into the ceiling, usually fluorescent.

Visual Comfort Probability (VCP): A mathematical system (usually computerized) used to produce a table or calculation of VCP values which evaluate the various glare sources and relationships in a room, as seen from the worst point in the room. The resulting percentage factor is that percentage of the room which is calculated to be comfortably free of glare. A value of .70 or better is considered to be desirable for usual comfort; a value of .80 or better is likely to prevent reflections in VDT screens.

Veiling Reflections: The reflected image of a light source on a task that reduces contrast and therefore, the visibility of the task. The name is derived from the veiled or film-like appearance a task appears to have as a result of the reflected light. The effect is best observed when looking at a piece of shiny paper, such as a page in a glossy magazine.

Work Plane: The plane on which work is being performed, usually horizontal and usually 30 inches off the floor.

Figure 16.3 (continued)

The Nature of Light

White light, whether daylight or produced by a lamp in a lighting fixture, is not all the same nor is it all of equal quality.

This section examines some of the more important characteristic of light.

Color

Kelvins: Light, as discussed here, is made up of different wavelengths varying from the very warm, long wave lengths to the very short, cool wavelengths. The color spectrum or color temperature of light, its *chromaticity*, is measured in Kelvins (K). Chromaticity values at the 3000 K range and below are considered warm, those at the 4000 K and above range are considered cool.

Color Rendering Index: While the Kelvin factor is a measurement of the temperature of light, it alone will not be an accurate indicator of the capacity or ability of the light to produce good color rendering. The International Commission on Illumination has developed a scale for establishing a measure for comparing the ability of a light to render accurate color. The scale is from 0 to 100, with a light source of CRI 100 or R 100 being the most accurate representation. The CRI index is a good index for determining how closely a light source will render ambient element colors, – but if comparisons are being made between two or more light sources, they must be of the same chromaticity (K).

A few years ago, lamps with high CRI index values were quite inefficient. Now lamps with high CRI's are as efficient as others of their type, but they are usually more expensive. This cost premium is due to higher cost phosphors used in their manufacture. Lamps are now available in the 80 to 92 CRI range, and at K's of 3,000 to 7,500. A CRI value of 90 meets most industry standards for color matching.

Reflectance

Reflectance is not something truly inherent in light, but in the surfaces which the light strikes – and off of which it bounces or reflects. Because our ability to utilize generated light is very dependent on the amount of light that is reflected off of surfaces, it becomes important to be able to rate the various surfaces reflecting the light. Again, this is done using a percentage factor. A perfect reflecting surface is 100 percent, or 1.00.

Figure 16.4 gives reflectance values for various materials and colors. Note that there are ranges for each material or color. These ranges principally compensate for sheen, and individual differences in interpretation of gray.

Most paint manufacturers show the LRV (light reflectance value) on the back of their paint sample chips, but usually only on those samples produced for the commercial interiors (architectural) market. Rarely does the LRV appear on sample chips found in residential paint stores or on chips produced for exterior paints. It will likely appear as "LRV 75%," etc. Most commercial wallcovering also indicate their LRV's, as will all commercial ceiling panels.

The more desirable ranges for visual comfort are listed in Figure 16.5.

Some of these ranges may seem quite low. In reality, they are not. This is in part due to the problems of glare, veiling reflections and contrast, which will be discussed later in this chapter.

Reflectance Ranges	
Material/Color	**Approximate Reflectance Range**
Acoustical ceiling	75-80
Aluminum – polished	60-70
Aluminum – brushed	55-58
Stainless steel	55-65
Light oak	25-35
Walnut	5-10
Marble	30-70
Clear glass	8-10
White	80-85
Ivory white	70-80
Ivory	60-70
Pearl gray	70-75
Light gray	45-70
Dark gray	20-25
Buff	40-70
Tan	30-50
Browns	20-40
Greens	35-50
Blues	35-60
Reds	20-40

Figure 16.4

Desirable Reflectance Ranges	
Floors	20-35
(Most carpet is less than 20.)	
Walls	40-60
Ceilings	60-80
Work surfaces	25-45
Office equipment	25-45

Figure 16.5

What We Want

When defining and designing a lighting system, we are trying to achieve certain criteria, some of which seem to be at odds with one another.

Proper Quantity: The most obvious goal is the proper amount of light for each of various tasks and areas. In the Basic Lighting Requirements portion of this chapter, some criteria were established which define light levels in terms of footcandles. These levels were predicated on lighting systems that are fairly unsophisticated, and must compensate with sheer quantity for a number of conflicting task needs.

Contrasts: We will deal with three elements of contrast. The first is contrast between surfaces, in particular the surface of the task being performed, i.e., the paper being read on the desktop, or a VDT screen and the source document. Eye movement from a very bright surface (white paper with writing on it could have an LRV of 60 to 70, while a walnut desk surface could have an LRV of 5 to 10), to a dark surface causes rapid eye fatigue. A similar, but more extreme example is the eye shift from a black on white source document to a VDT screen with light copy on a dark background. How we adjust for these contrasts is a part of our contrast concerns.

The second form of contrast which concerns us is the contrast between viewing areas. Assuming that lighting levels have been reduced to the B Illuminance Category, or the 5 to 10 footcandle range, what should be done about the surrounding areas where a D Illuminance of 20 to 50 footcandles is needed? More importantly, what about the existing 75 to 100 footcandles already in place or planned? Eye fatigue in these surroundings can lower worker efficiency in just a few hours – far fewer hours than a normal eight-hour shift.

The final contrast concern should now be fairly obvious: the contrast between the material being read (studied, calculated, analyzed, etc.) and the surface on which it is printed. If the lighting levels are in the 20 to 50 footcandle range suggested for VDT work, our ability to rapidly and accurately read even the good contrast of black print on a white document can be severely reduced by veiling reflections from a misplaced source of light on the task surface.

The best solution would be to increase the contrast of the task document, without raising the brightness of the document more than about five times that of the VDT. The general rule is that every 1% reduction in task contrast, due to veiling, will require a commensurate illumination increase of 10% to 15%.

What We Get

We are beginning to develop a feel for the type of lighting we are trying to achieve, but in reviewing some of the elements of light, surfaces, and quantity, we have begun to recognize other concerns such as glare, veiling reflections and shadow.

Glare

Although glare has long been an obvious concern from windows and some reflections, it has not been until recently that we have become seriously concerned about it from ambient, ceiling-mounted troffers.

Glare from windows can be controlled with horizontal blinds and light-weight draperies, and if the building is modern enough, through the use of heat-absorbing and reflective window glass. In all events, the desirable solution is to achieve an equalized level of light, while still permitting outdoor viewing without rendering the view gloomy with too-dark a glass.

Glare from fixtures is a different problem. The concern is occurring more frequently due to the more frequent use of VDT screens. As we look straight ahead at a VDT screen, our normal peripheral vision will be aware of light glare within a 45 degree field of vision. If we have a ceiling height at 8'-6" and our eye plane is at about 4', 45 degrees of peripheral vision would become aware of any ceiling-mounted light fixture glare 4'-6" in front of us, and beyond.

The VDT screen itself becomes an even greater offender, as it picks up reflective glare behind and to the sides of the viewer because of its convex shape. This on-screen glare reduces contrast between the screen image and the screen background.

The screen also tends to reflect aberrant images of other bright room elements, such as mirrors, pictures, etc. The screen itself will often become a mirror reflecting the operator's image and reducing contrast.

There are no universal solutions to the problems of glare. One area of a room may have little or no direct glare on a VDT screen. Another VDT screen placed in the same orientation, but a few feet away may receive direct or reflected glare from an unanticipated source. Some of the better solutions seem to be the elimination or reduction of natural outside lighting, coupled with the use of indirect ambient lighting, good vertical illumination (light reflected off of walls or partitions), and the use of individually controlled task lighting. The task lighting may need to be on an adjustable arm so that it can be individually aimed and equipped with a dimming device.

Veiling

Veiling reflections are another form of glare that stem from the task surface itself. The task surface reflects light rays in such a way that we observe them as light, and this substantially clouds or veils the surface. This problem is readily observed when trying to read glossy magazines or looking at photographs. The condition may also occur, more subtly, in non-glossy surfaces. The most offending light sources for veiling are the zone from 0 degrees vertical to 30 degrees from vertical on a horizontal task. The solutions for veiling are much the same as for glare.

Shadows

Surprisingly, shadows are one of the lesser problems with which we must be concerned. This is due to the large size of most ceiling-mounted fixtures, the overlapping light patterns from the fixtures, and the reflectance value of the ceiling panels. However, task lighting located under an overhead systems furniture storage unit may generate a shadow problem—along with glare and veiling reflectance problems.

Solutions

Now that most of the principal measurement criteria and technical problems related to lighting have been reviewed, it is time to consider some of the potential solutions. We will be looking at six alternative solutions. There could be many more solutions which might involve various mixtures of these six. Examine as many as possible, but do try to achieve a better result than Solution 1.

1. *Uniform Ambient Direct Troffer Lighting Reflected Ceiling Solution*

   This is the most common solution in use today, thirty years ago, and probably for the next few years. It involves installing 2' x 4' ceiling troffers 8' to 10' on center, using a standard plastic prismatic diffuser. There are three reasons for this solution's continued popularity. Cost is first. Most office space is leased "as is," or on the basis of a "build-out" with a tenant work letter which provides for a building standard light fixture on the basis of one fixture for every 75 SF to 100 SF of rentable area. The landlord's standard fixture is most likely to be whatever is the cheapest he can provide given the competitive market. Not only is this fixture cheap, but it is efficient, producing the most light (though of poor quality), with the fewest fixtures.

   The second reason is ignorance. The fact that many interior designers are not well qualified to design a quality lighting system – coupled with the general perception of the public and most employers that "light is light," does not result in much incentive to the interior designer to try to justify the premium cost of an upgraded lighting design.

   The third reason is inertia. It is a lot easier to go along with what has been done in the past, what the landlord is offering and what the tenant is used to, than to go through the difficult and costly effort to explain, design, cost justify, pay for, and install a quality system.

2. *Indirect Solutions*

   This solution usually calls for the suspension or furniture mounting of specially designed fixtures which direct their light up against the ceiling. Either of these may use wall-mounted fixtures to provide vertical lighting reflected off of walls and partitions.

   These systems can provide good quality uniform lighting for most tasks. The lighting fixtures must be suspended a minimum distance below the ceiling – one-fourth the distance between the fixtures. In an 8'-6" high room with the fixtures about 7' from the floor, the room can develop a cluttered feeling, as well as nonuniform reflectance patterns on the ceiling when the fixtures are spaced more than six feet apart. This is also true of furniture-mounted systems and some of the larger HID free-standing units. HID units often provide nonuniform color as well as nonuniform patterns. Indirect fixtures require considerably more cleaning maintenance because of their involvement in the air circulation patterns and their closer proximity to areas of dust generation. These systems, depending on the quality of their design, may require some localized task lighting as fill-in.

3. *Task and Nonuniform Solutions*

   These solutions involve the nonuniform placement of ceiling troffers so that they are in close proximity to workstations. Ideal placement would be over each end of the workstation work surface, running parallel to the ends of the work surface. Once in place, this solution is quite inflexible. It works fairly well in private offices, where furniture placements tend to remain fixed, particularly when parabolic louvered fixtures are used. Some specifiers feel that this solution alone will correct veiling reflection problems. The author's experience is to the contrary due to the task lighting fixture usually being mounted in a fixed position under a systems furniture storage unit. This places the lighting fixture at about eye level, projecting the light down onto the work surface. The intensity of the light and its close proximity to the work may produce undesirable glare and veiling.

4. *Task/Ambient Solutions*

   These solutions utilize a combination of ambient lighting augmented by specific task-related fixtures. Often the ambient portion is through the use of furniture-mounted ambient fixtures. These fixtures will be on the tops of systems furniture storage units or on the top edges of systems panels. The development of high-output, compact fluorescent lamps has generated increased interest in the development of ambient-indirect fixtures as an inconspicuous part of the systems furniture. This application has the added benefit of placing the ambient system along the periphery of the workstation. Fortunately, lamp life and light output of the high output compacts compare quite favorably to that of the standard fluorescent 20,000 hours and 3,150 lumens for the standard 40-watt T12 lamp. Task-oriented fixtures, usually mounted under systems storage units, often have disturbing shadow, glare and veiling reflectance characteristics. Task/ambient systems have some very favorable aspects, but have been very difficult to install with consistent results.

5. *Ambient Solutions*

   General ambient systems, using state-of-the-art parabolic-louvered fluorescent fixtures, seem to be one of the better solutions. By adding the appropriate lamps, excellent color rendering can be achieved. Most glare related to fixtures will be reduced if not eliminated. Veiling reflection will be reduced, but not eliminated. To eliminate veiling, polarized lens correction would have to be added. Any daylight glare problems must be dealt with through window treatment. But without special controls, no daylight harvesting is achievable. Areas that are non-task-oriented should be dealt with on an individual basis to provide special effects and reduce or increase illuminance. Ambient systems using parabolic-louvered fixtures are superior systems, providing better-than-average lighting, but without much in the way of energy savings or conservation.

6. *Ambient/Controlled Solutions*

   Fortunately, systems are now available which can produce solutions to just about all of the concerns we have reviewed so far in this chapter. They can also render very substantial reductions in energy costs during normal task-operating times, and after hours. These are the ambient/controlled systems which utilize currently available technology from different sources, and can produce very high quality and responsive systems. These systems are somewhat more costly to install, but have good ROI's (return on investment), and continue to provide superior, flexible and responsive illumination at well below the operating costs of other systems.

   These systems consist of high-quality parabolic-louvered troffers, possibly coupled with polarizing lenses. The luminaires are fitted with dimming, solid-state ballasts. These systems may utilize computer-controlled light levels and scheduling which cannot only can turn fixtures on and off, but can also reduce lighting levels to as low as 15% to accommodate to the most critical task operations, as well as for cleaning, general maintenance or stocking. These dimming ballasts can be interfaced with light-sensing devices installed to harvest daylight and to automatically regulate the ambient system to pre-programmed lighting levels. Inexpensive motion or infrared occupancy sensors can be set to either turn lights off or to reduce lighting levels to pre-programmed levels when specified areas are unoccupied.

This type of system is not only nearly glare-free, but may substantially reduce veiling reflections. It can also improve contrasts, thereby permitting substantially reduced task-related lighting levels; and can harvest natural daylight, reduce output or turn off lights in unoccupied areas, extend lamp life, and reduce energy consumption and ballast heat, even when at full capacity.

These types of systems are not only superior to the "brute force" systems, but are more economical to maintain and to operate. They should recover their premium cost in from 1-1/2 years to 3 years depending on local energy rates and the size and sophistication of the installed system.

The technology of the fixtures, lamps, lenses, and control devices will be discussed in subsequent sections of this chapter.

Fixture Types

Fixtures are designed first as a housing for a particular type of lamp. The most common lamps employed in typical office space are first fluorescent, then incandescent, and finally, some of the HID types. Of these three, the incandescent and the tungsten halogen (which is an incandescent) are essentially used as accent lighting to highlight a particular object, area or feature. It is not advisable to locate incandescent fixtures directly over a seating area due to the heat incandescent lamps generate. Compared to the fluorescent and HID's, they are most inefficient and should not be considered for normal ambient office lighting. Some lighting designers have had some success in using mercury vapor and metal-halide lamps. Although many of these still use a basic "can" design, either surface-mounted or recessed, some are now using a troffer design, and still others are using a pendent inverted mushroom design. The pendent type fixtures hold great promise for extended use in ambient systems.

Depending on the construction of the fixture or troffer and the "shielding media," most fluorescent fixtures have been rated for a VCP (Visual Comfort Probability)* range and for efficiency. Luminaire efficiency is simply the lumens given off by the fixture, in an operating condition, divided by the total lumens given off by the fixture's lamps.

VCP range is a measure of comfort. The VCP of a fixture is based on a series of computations, taking into consideration such elements as the luminaire shape and size, room size, room height, reflectance, various angles of view, and most importantly, the glare given off directly by the luminaire.

An index of 0 to 100 is developed, and a fixture is assigned a VCP factor for a given set of factors. As the factors change, the fixture's VCP will probably change. A VCP factor of 70 is expected to be acceptable. In other words, 70% of the room is comfortable, while 30% is not. Since VCP should be calculated for all portions in a room, this means that 30% of the space within the room is uncomfortable. If we assume that VCP is valid, there is a lot of room for improvement in the design of lighting fixtures.

Fluorescent

The most common fluorescent fixture is the 2' x 4' with four 40-watt lamps. This fixture uses two ballasts. Usually one of the ballasts is wired to the two outermost lamps, with the second ballast wired to the two inner lamps. Most commonly, it is fitted with an acrylic diffuser. This fixture also comes in a 2' x 2' size and a 1' x 4' size, both with single ballasts.

Fluorescent fixtures are now available with a variety of "shielding media" (lenses or louvers). These have VCP's in the 45 to 99 range. Unfortunately, they likewise have an efficiency range of 25% to 70%. A few comparisons are shown in Figure 16.6.

The only shielding material from this list that effectively eliminates veiling reflection is the polarizing lens. Coupling a parabolic louver with a polarizing lens produces a very high quality luminaire.

Systems

Most office lighting is fluorescent due to its operating efficiency and low fixture cost. Therefore, four of the most common fluorescent office lighting systems, along with their advantages and disadvantages, are reviewed in the following section.

Floor-Mounted and Furniture-Integrated Task/Ambient

This system uses a combination of furniture-mounted fixtures projecting their light upwards on the reflective ceiling to achieve a low luminescence ambient lighting. These luminaires might be augmented with floor-mounted, upward-projecting fixtures also bouncing light off of the ceiling. The floor-mounted fixtures are often metal-halide. Usually task lighting is necessary to upgrade the light to a working level.

Advantages
- Flexible.
- Few, if any, ceiling-mounted fixtures required.
- Ease of maintenance.
- Better acoustics, due to lack of ceiling penetrations.
- Possibly interesting lighting patterns.
- Possible energy efficiencies.

Disadvantages
- Special furniture requirements.
- Higher cost lighting fixtures, possibly higher cost system.
- Higher luminaire dirt depreciation.
- Complex lighting system design.
- Relocation of systems furniture may upset lighting coverage.
- Task lights can produce shadows, reflectance off of working surfaces, and veiling reflections.

Integrated Ceiling Baffled System

This is a complete system of metal baffles, suspended about 12" below a reflected ceiling. Surface-mounted fluorescent fixtures are spaced in continuous rows above the baffle system. The result is a light ceiling which can provide a low-glare, high luminosity.

VCP Ranges Versus Efficiencies for Fluorescent Fixtures		
Shielding	**VCP Range**	**Efficiency**
Small cell parabolic	99	35-45
Deep cell parabolic	70-90	50-65
White metal louver	65-85	35-45
Polarized lens	60-70	55-60
Diffuser lens	40-50	40-60

Figure 16.6

Advantages

- Architecturally interesting.
- Good acoustics.
- Uniform light dispersal.
- Easy lamp maintenance access, depending on baffle construction.

Disadvantages

- Assuming a 9' ceiling, the bottom of the baffles will be at about 7'6", which detracts from the architectural values.
- Expensive.
- Requires a high ceiling.
- Difficult zoning control of light levels.
- Very inflexible.
- Very difficult and costly to add floor to ceiling partitions.

Ambient Pendent System

This system consists of rows of suspended fixtures about 14" from the ceiling, reflecting their light off of a reflective ceiling. This type of system has a modulated lighting tone. It can be quite architecturally interesting.

Advantages

- Permits both initial and relocation flexibility of furniture.
- Easy maintenance accessibility.
- System is considered glare-free.
- Low veiling reflection problems.
- Can be interesting architecturally.

Disadvantages

- Very difficult to build or relocate floor-to-ceiling partitions due to end-to-end configuration or rigid patterns of mushroom type fixtures.
- Reflected ceiling must be of high quality.
- Frequent complaints of insufficient lighting at task.
- Fixtures are quite expensive and must be handled very carefully.
- Unless the system is far enough off of floor, the architectural statement may be negative.
- Frequent complaints of lighting being "dull."

Recessed Ambient Ceiling Troffer System

This system employs a variety of 2' x 4', 1' x 4' and 2' x 2', 4' x 4' recessed fixtures. There have even been some 5' modules available. They may have diffuser lenses, parabolic louvers from 5/8" x 5/8" to 8" x 8", polarizing lenses, baffles or translucent nondiffusing lenses. The recessed fluorescent fixture is the workhorse of office lighting and is, by far, the most commonly employed. It is also the system on which the greatest amount of technical research is focused, both from the standpoint of energy conservation and lighting efficiencies.

Advantages

- Low fixture and installation cost.
- Can be employed in low or high ceiling areas.
- Often fixtures can serve as air handlers.
- High flexibility in terms of fixture orientation and relocation.
- Relatively easy to install or relocate floor-to-ceiling partitions.
- Greatest variety of fixtures from which to select.
- Can be easily task-oriented.

Disadvantages

- Ceiling architecture usually cluttered.
- Possible strong glare problems.
- Possible strong veiling problems.
- Often poor acoustics.
- Often poor energy efficiency.

Lamps

Incandescent

Although there is a great variety of lamps available with any number of variables in output, efficiency, color, size, shape, wattage and function, they can be classified as three basic groups: *incandescent, fluorescent* and *high intensity discharge (HID)*. In discussing lamps, references were made to Kelvin temperatures (K) and to Color Rendering Indices (CRI). The standard base of reference in the lighting industry is a clear day, noon sun, which will have a:

- Kelvin factor of 5,300 to 5,800
- CRI of 100.

The data shown in Figure 16.7 applies to the more commonly used lamps.

Within this group are the well known household incandescent light bulbs, designated by the letter A and ranging from A-15 (15-1/8th inches of bulb diameter) and 15 watts to A-21 (21-1/8th inches of bulb diameter) at 150 watts. Above 150 watts the incandescent lamps are designated as *PS*, describing the bulb's shape (Pear Shaped) with wattages to 1,000. Most of the A and PS lamps have a K (Kelvin) of around 2,900 and a CRI (Color Rendering Index) of 100. Lamp life for GS (General Service) lamps is 750 to 1,000 hours and up to 2,500 hours for ES (Extended Service) lamps.

A second group of incandescents are the PAR and R lamps. PAR stands for parabolic, and R for reflective. These lamps range from 30 to 500 watts. Again, the number following the letters describes the diameter of the lamp in eighths of an inch. Most PAR's and R's have K factors, CRI's and lamp lives about the same as the A and PS lamps.

The third group of incandescents are the tungsten halogen lamps. These lamps range from 20 to 1,500 watts and have lamp lives up to 3,000 hours. K factors are about 3,000, with a CRI of 100. Many tungsten halogen lamps are used in PAR lamps.

Incandescent Characteristics	
Incandescent	**Ranges**
Color Temperature	about 2,900
CRI	100
Wattage ranges	7 to 1,000
Efficacies (lumens/watt)	15 to 25
Lamp life (hours)	750 to 2,500

Figure 16.7

Fluorescent

Fluorescent lamps come in a variety of configurations, shapes and wattages. There are many different lamps available, with new iterations coming to the market constantly. This review is limited to one or two lamp shapes and sizes.

The most common fluorescent lamp is the T-12 or T-8, 40 watts, 48" long. As with the incandescent, the numeric designations 12 or 8 represent the diameter of the lamp in eighths of an inch, a T-12 being 1-1/2 inches in diameter.

Most manufacturers now produce 32, 34 and 35 watt lamps to fit the standard 4' fixture. Although they do represent a savings in energy, they have a commensurate drop in efficacy. Most manufacturers also produce a variety of so-called high output, as well as color corrected lamps. Color renditions are now excellent, with broad K temperatures and CRI's as high as 92 for some specialized chroma-corrected lamps.

The industry is responding to an increasing demand for lamps for the 2' x 2' fixtures which are increasing in popularity due, in part, to their non-linear ceiling appearance. These lamps now come in a U-shaped T-12, with a 3" or 6" leg, and a compact T-5 lamp with wattages of 7, 9 and 40. These lamps are now developing lamp lives in the 15,000 range and have most of the color temperature and CRI factors of the standard 48" lamps.

Because fluorescent lamps are very temperature-sensitive, they are not recommended for outdoor use in cold climates.

High Intensity Discharge (HID)

This is a group of three different lamps, all of which operate on similar theories and have similar construction. They consist of two glass envelopes, one inside the other. The outer envelope protects and insulates the inner envelope from temperature loss while operating. There is a main electrode which begins the lighting process by developing very high temperatures. An arc is initiated from this electrode. The inner shell, lined with metallic coating, completes the arc.

Mercury Vapor

Mercury vapor lamps have been used for some high intensity interior lighting, but their poor color rendering and low efficacies mean that they are usually a poor choice.

Fluorescent Lamp Characteristics	
Fluorescent	**Ranges**
Color Temperature	2,700 to 7,500
CRI index	52 to 92
Wattage ranges	5 to 220
Efficacies (lumens/watt)	55 to 90
Lamp life (hours)	9,000 to 30,000

Figure 16.8

Metal Halide and Metal Arcs

These are very efficient lamps and are a good "white light" source, especially for large, high-ceiling spaces; down lighting; or indirect lighting.

High Pressure Sodium (HPS)

This is the highest output lamp, with efficacies of 70 to 140 lumens per watt. Because of their high efficiency and wattage availability, they are frequently used for outside lighting, where color rendition is not of prime importance. Currently, standard high pressure sodium lamps are decidedly biased to the yellow in rendition, with little blue or cooler colors. Some HPS color-corrected lamps lose their color rendition quickly, even though they have extremely long lamp lives. Some technological improvements may be forthcoming in HSP ballasts, which may correct this problem.

Figure 16.9

Mercury Vapor Characteristics	
Mercury Vapor	**Ranges**
Color Temperature	about 3900 K
CRI index	about 50
Wattage ranges	40 to 1000
Efficacies (lumens/watt)	30 to 55
Lamp life (hours)	12,000 to 24,000

Figure 16.10

Metal Halide and Metal Arc Characteristics	
Metal Halide and Metal Arc	**Ranges**
Color Temperature	3,600 to 4,300 K
CRI index	65 to 85
Wattage ranges	75 to 1,500
Efficacies (lumens/watt)	80 to 115
Lamp life (hours)	6,000 to 20,000

Figure 16.11

High Pressure Sodium Characteristics	
High Pressure Sodium	**Ranges**
Color Temperature	1,900 to 2,300 K
CRI index	22 to 65
Wattage ranges	35 to 1000
Efficacies (lumens/watt)	70 to 140
Lamp life (hours)	15,000 to 24,000

Ballasts   Ballasts are an integral device in all fluorescent and HID fixtures. They have two functions, first to provide the voltage to the filaments (which results in electrons being "boiled" off the filament); and second, to control and regulate the flow of current through the lamp. The current, in turn, excites the gas, which begins to emit ultraviolet light, causing the phosphorous coatings on the inside of the tube to begin to glow, emitting visible light. Once the lamp has reached its targeted output level, the ballast must cut back on the power to provide a constant emission of current to maintain the proper electron discharge, and therefore, a consistent light level.

Most fluorescent ballasts are magnetic type, consisting of a core and coil insulated by dielectric tars and, until recently, PCB's. (If the ballasts in your existing spaces were installed prior to 1979 it is possible they could contain PCB's.)

There are a number of electronic ballasts available. Most are really hybrids consisting of part magnetic and part solid state. These units have had a fairly high failure rate. They have also produced undesirable amounts of electromagnetic radiation and harmonics.

There are also fully electronic ballasts, some of which can provide full dimming capabilities from 100% output to 5%, which emit very low electromagnetic radiation and virtually no harmonics. The fully electronic ballast is vastly superior to the magnetic type or hybrid types, for a number of reasons. The best quality electronic ballasts can provide the following advantages over the magnetic or hybrid:

- High degree of reliability, energy efficiency; a reduction of 20%-30% energy consumption, with equal illumination from equal lamps;
- Elimination of power line harmonic distortions and electromagnetic radiation interferences;
- Elimination of ballast noise which was caused by vibration of core elements in magnetic ballasts;
- Elimination of lamp flickering by increasing the AC cycles from the standard 60 Hz to 20,000 to 30,000 Hz. This frequency is so high that neither the eye nor the brain are sensitive to the light cycling. The eye may not be aware of the cycles in the 60 Hz range, but the brain is. Additionally, the 60 Hz range can cause a strobing effect, particularly in some VDT's.
- Some electronic ballasts can provide properly controlled dimming. Some hybrid types can also provide dimming, but will substantially reduce lamp life.
- Increased lamp life (by 20%-25%) by ramping the heating of the filaments over 500 milliseconds, rather than providing a full voltage charge at one instant.
- They will work effectively and efficiently with computer controlled EMS (Energy Management Systems).
- Reduced heat output of the ballasts, during full output, by 20%-25%, and when dimmed, a commensurate reduction in heat output, with consequential reductions in air-conditioning needs.
- These types of ballasts also have an inherent "brown-out" protection.

Energy Management Systems (EMS)

Energy Management Systems have been available for the last decade. Until the development of the true electronic ballast, most EMS applications were circuit related; that is, they turned various circuits on and off primarily at pre-scheduled times. Some systems were able to utilize infrared sensors to turn off circuits when there were no people present. By in large, EMS was only marginally effective until the introduction of the electronic ballast. With the development of the dimming electronic ballast, major control of lighting systems may now be achieved, with lighting-related energy cost reductions as high as 85%.

The principal elements of such a system would consist of:

- The EMS computer; or a number of microcomputers called *Master IC's* or *Inter Coupler's*,
- Infrared or motion occupancy sensors, to either turn off individual or groups of ballasts or to reduce lighting to pre-determined levels,
- Photo-electric sensors to provide sunlight harvesting and to adjust lighting levels to proper interior/exterior contrast levels. These devices can modulate up and down to provide exact, pre-determined, task-oriented lighting levels.
- Local manual variable power rotary or slide dimming switches to permit off-hour space utilization or temporary changes in lighting level requirements.

If a planned space is large enough to produce a positive return on investment on the cost of an Energy Management System, the company and its consultant should give such a system serious consideration. A well engineered and managed EMS's ability to adjust to various light level needs or employees' desires may be substantial, in terms of improving productivity and other economic benefits.

Maintenance

Maintenance of lighting systems is fairly straightforward, but like carpet maintenance, is usually not practiced until there are obvious problems. There are two major sources of light deterioration, with dirt and dust being by far the most deleterious. One of the most efficient ways of reducing dust and dirt in fixtures and on lamps is to be sure that the air handling system receives regular filter replacement. The predominant source of dirt, grease and dust is distributed through the air handling system.

Even with the best of filters, some lamp and fixture maintenance is required. If the facility's air handling system is not the best, or if the windows are open a good bit of the time, an annual cleaning is in order. On the other hand, if the air handling system is in good order and well maintained, cleaning needs to be no more often than when lamps are changed.

Lamp Changing

Spot Replacement

There are two schools of thought on lamp changes. The more common, if not the best, is to replace lamps when they burn out or when enough have burned out that the space begins to look bad and workers are complaining about dark spots or excessive flickering. If this is the company's practice, at least make sure that a periodic cleaning of the reflective surfaces of the fixture is performed with a damp cloth when lamps are changed. Lenses should be damp wiped, small cell parabolic louvers should be dipped in a mild detergent and air-dried. If a Spot Replacement program is in place, all of the lamps in the fixture should be replaced at one time.

Group Replacement

The second school of thought is to replace all lamps on a programmed basis. The labor cost for this program is 10% to 20% of the labor cost for Spot Replacement, and the savings in the labor cost is, by far, greater than the additional lamp costs. Fixture, louver and lens cleaning would be part of a group replacement program. The timing of a group replacement program will depend on the number of hours the system operates per year, and the efficient life of the lamps in use.

Fluorescent lamps typically have a life of between 9,000 and 24,000 hours. These are theoretical hours and do not take into account lamp lumen depreciation, which will run about 15% to 10% (less with high quality electronic ballasts). The average rated life of quality light for most fluorescent lamps used for ambient lighting is about 10,000 to 15,000 hours. An average system will operate about 2,500 to 4,000 hours per year, which indicates a group replacement every 2-1/2 years to 4 years.

Life Safety Codes

We are all aware that current life safety codes require emergency and exit lighting in all occupied spaces. The most common is a system which will provide 1 footcandle for 1-1/2 hours, with little or no interruption. This lighting must be at exit access areas, which includes all corridors, ramps, stairs, aisles and passageways leading to an exit. In addition, illuminated EXIT signs are required at all designated points of egress. The company's interior designer, architect and building landlord should be able to provide guidance in this regard.

ASHRAE (Codes)

Since the energy crisis of the mid 1970's, many sections of the country have adopted "energy power consumption" codes. Many buildings are designed with energy budgeting (conservation) as a part of the building standard. Ask the landlord what the energy standards are for the planned space.

The American Society for Heating, Refrigeration and Air Conditioning Engineers (ASHRAE), in conjunction with the Illuminating Engineers Society (IES), has established procedures for calculating Energy Power Budgets. These are gross budgets which describe permissible power in terms of watts per square foot. They are not design criteria per se. They merely set the maximums for power usage. The lighting engineer or designer is at liberty to design within the prescribed usage allowances.

Site Lighting

Site lighting may not be a consideration in designing and building out planned facilities. However, the actual and perceived safety of the company's employees is. Part of the original assessment of the building had to do with employee safety and exterior lighting. Just because the building space ultimately satisfies all of the company's needs does not mean that the exterior lighting is satisfactory. The matter of lighting operating hours, lighting intensity, and areas of illumination should be carefully reviewed, in the evening, at night and early in the morning. The person or persons doing the surveying should do so from the point of view of all types of potential building users and visitors. For example, men and women may have quite different views of what kind of lighting is adequate for security. The observations and concerns of all individuals, including those who are physically challenged, should be considered.

Architectural Lighting

Today, many buildings are architecturally lighted. If the company has secured an exterior sign and is permitted by code and the landlord to display it, consider lighting it. A review of the building architectural lighting may also provide some security lighting. Often small modifications to the architectural lighting can provide additional security lighting without detracting from the architectural design.

Chapter 17

ACOUSTICS

Introduction

On first considering *acoustics*, we may immediately think in terms of distracting noises, usually from another area. In a very real sense, we are right in so doing. Rarely, however, do we consider that *we* are a part of the problem. Each of us is a generator of noise—insofar as someone else may be concerned. The primary function of an office environment is communication. Even the most mathematical or research-oriented job functions rely primarily on oral communication. We provide and receive all sorts of instructions and information by speaking to our fellow workers and by being spoken to by them, whether face-to-face or over the telephone.

Spoken communication is the most common work function we perform. Not only is speaking a work function, but for many of us, it is one of our most frequent forms of recreation—What the ball game scores were ... Where the best buys are ... Who is seeing whom. All of these forms of spoken communication are noise generators. The problem we face is how to participate in the spoken communication process and yet not distract others who are not directly involved. The second part of our consideration is how to deal with all of the other ambient sounds bombarding our ears and those of our fellow workers.

While a great deal can be done to improve acoustics in the office environment through careful understanding of the fundamentals of sound and its nature, there will still be a great number of compromises. However, never forget that communication is the singularly most important objective of our office. Enhancing that objective is of paramount importance.

The Characteristics of Sound

Elements of Sound

To examine the elements and nature of sound, a basic understanding of some of the fundamentals of sound is needed. We will try to do this without becoming involved too deeply in the technical aspects. The first element is that sound needs some sort of medium in which to exist and travel; solids, liquids, and gases will all transmit sound. Because sound is a form of vibrating energy moving in concurrent waves, its energy is rapidly consumed in vibrating whatever medium is transmitting it. Its energy is quickly converted to heat and decays into silence.

There are three basic measures of sound:

Velocity: how fast it travels,

Frequency: how high or low-pitched it is, and

Power: how loud it is.

Velocity

Sound travels at about 770 miles per hour. This is the so-called sound barrier which jet aircraft pass through, creating a loud bang as they do so. This translates to about 1130 feet per second at sea level. This means that sound emissions which are fairly close by, and even though not very loud, will reach our ears almost instantly.

Frequency

When a sound is generated from any source, it takes the form of a series of vibrations in waves, which are basically spherical in shape, like the layers of an onion. The more rapid the vibrations or cycles, the higher pitched the sound. The less rapid the cycles, the lower pitched the sound.

The frequencies are measured in cycles per second and are defined in Hertz (Hz). One cycle per second = 1 Hz. Although audible sound frequencies extend over a very wide range from near 20 Hz to in excess of 12,000 Hz, the usual spectrum is from 60 to 8000 Hz and is divided into eight octave bands of 63, 125, 250, 500, 1000, 2000, 4000, and 8000 Hz.

When sound energy strikes a wall:

1. Some reflected sound adds to the sound field in the room;
2. some sound is absorbed and dissipates as heat; and
3. some transmitted sound energy propagates through the walls, floors or ceilings.

Power

The third characteristic of sound is loudness or power. Because sound is a form of energy and because it radiates out from its source in a spherical form always enlarging, it begins to lose energy (and loudness) rather rapidly.

Technically, the intensity is inversely proportional to the square of the distance from the sound source. Simply stated, the greater the distance from the sound source, the lesser the intensity. As a result, we are interested in the difference between sound intensity at its source and at the point of its perception.

We measure the intensity of sound in decibels. A decibel (dB) is a measurement of what the human ear perceives as the intensity level per square unit of a sound as compared to the intensity of the sound at its source. The decibel is a logarithmic calculation; therefore we cannot add decibels together without adjustment or unless we do so using logarithmic calculations.

In order that we may have a scale of relative decibel levels, the chart in Figure 17.1 will help to set some standards of comparison. Remember the decibel levels shown are those perceived by a human ear and not the levels at the source of the sound.

Nature of Sound

Paths

In many ways, sound travels in paths much like a directed stream of water. It will flow through almost any opening, in any direction. It will flow upstream in an air supply duct, not as easily as downstream through a return air grille or duct; but nonetheless it will travel upstream. It will travel through a crack under a door or through a joint between two systems furniture panels, through a hole in a wall or ceiling. Not only will sound flow or *diffract* over and around some barriers, it will reflect or bounce off of barriers such as walls, ceilings, and floors.

Sound will also pass directly through some barriers. For this reason, a number of basic measuring ratings, indices or coefficients has been developed in order to measure the ability of barriers to impede sound transmission, or to measure the ability of materials to absorb sound, and how long sound lasts.

Reverberation and Duration

Reverberation time is the measurement or awareness of how long a sound persists after it is generated. If the reverberation time of a particular voice is extremely short, which might be good for keeping kids quiet, we might not be able to understand what was being said to us. On the other hand, if the reverberation time is too long, the sound of one voice would begin to overlap the sound of another voice. When we consider duration and reverberation, we should examine their positive and negative values. For office space purposes, we would like to achieve rather low reverberation times; something in the 0.3 to 0.8 second range. A moderate-sized corporate auditorium might try to achieve a 1.2 to 1.6 reverberation time. A concert hall might wish to achieve a 1.6 to 2.2 reverberation time, while a major cathedral might try to achieve a time of 1.8 seconds or longer. For an office, a

Sound Source Comparisons		
Sound Source or Example	**Comments**	**dBA**
Commercial jet aircraft	physical discomfort	130
Gunshot	physical discomfort	130
Ambulance siren at 100 ft.	all other sounds excluded, deafening	120
Major city street noise	loud, distracting, tiring	90
Operating factory	difficulty conversing	80
Typewriter, impact printer	distracting	70
Average large office	loud conversation, moderate distraction	60
Private office	comfortable level	40
Quiet conversation	intent listening	30
Soundproof space	small sounds become distracting	10
Threshold of human hearing		0

Figure 17.1

reverberation rate below 0.2 would not be bright enough and would give the space a dead feel. Above 1.0 seconds could give a cluttered sound feeling. A simple example of reverberation time is to remember the reverberation in an enclosed swimming pool or a fully tiled bathroom. In both instances, reverberation time is relatively extended.

Reverberation time is a function of room volume and the capacity of the room to absorb sound energy. Reverberation time is the amount of time it takes for a given sound to decrease 60 dB after the source has stop emitting the sound.

Reflectance

Too often designers do not consider the problems of noise reflectance until they have created a problem they must then try to solve with some sort of noise reduction solution. Some recognition of the conditions that create undesirable reflectance should be considered during all phases of the design stage.

When selecting materials in sound-sensitive areas, consider that ceramic tile or masonry walls, windows, glass partitions, metallic wall finishes, banks of metal file cabinets, painted metal systems furniture panels, plastic light lenses, etc. all act as highly reflective noise surfaces. The total capacity of carpeting and acoustical ceilings to dampen sound can be overwhelmed by too much reflective surface in the perimeter. A ceiling with a high *noise reduction coefficient* (NRC) but with an excess of plastic lens light fixtures can have its effectiveness reduced by 30 percent. When the plastic lighting lenses are placed so as to provide the most efficient lighting (at roughly 45 degrees from the task surface), the effectiveness of the sound absorbing materials, in the same area, can be reduced by as much as 50 percent. The anomaly is that the angle of maximum efficiency for the light fixture may well be the worst angle for sound reflectance.

Transmission of Sound

Sound can and will pass through a physical barrier such as a wall or ceiling. It can also be conveyed, or transmitted by a solid mass, such as a floor or wall over large distances. Once we recognize these facts, we must plan on when and how to avoid objectionable sounds, or if unavoidable, how to control them.

(TL) Transmission Loss

Transmission loss (TL) is a characteristic of a sound barrier measured by the difference between the SPL (sound pressure level) at the sound source; minus the SPL at the point of perception. The TL varies with the sound frequency (HZ) of the sound and the mass of the barrier. It is known that a given TL will increase about 6 decibels (dB) for each doubling of the barrier mass. This assumes a single homogenous barrier mass, such as a single sheet of gypboard.

Example:

Weight of mass in psf.	5	10	20	30	40	80	160
Transmission Loss in dB of 400 Hz.	33	39	45	51	55	57	63

There are a number of factors which distort this rule of thumb. A barrier of multiple layers, such as two layers of 1/2" gypboard laminated to each other, but of equal mass to a homogeneous 1" layer, will have a superior TL.

The net effect is to think of sound transmission control in terms of mass. Think in terms of sound absorptive materials, limp — not elastic. Think in terms of cavities or air spaces, of double walls or staggered stud walls, each with insulation. A wall of unbalanced masses (more on one side than the other) is more effective than a balanced wall of equal mass. A doubling of air space within a barrier, can have a greater effect on TL than pure mass. The problem with this is that the smallest effective cavity of air space only (without any insulation) is 2″, and thickening a wall to provide a 12-inch air space is not practical.

A transmission loss is determined at a specific frequency. Different frequencies will produce different transmission losses; consequently TL's are often shown as having a spread in order to span the frequencies from 125 Hz to 4000 Hz. A single TL becomes a fairly useless measurement for real world work. Because of this a different system of transmission loss was developed utilizing a series of TL tests from 125 Hz to 4000 Hz, all for the same barrier. The tests are extrapolated to a standard contour which smooths or modulates out unusual dips or bumps in the curve and down-rates the contour accordingly. This laboratory test curve is used to establish sound transmission class (STC). The STC rating of a sound barrier is what concerns us in facility design.

(STC) Sound Transmission Class

This is the first of the two most important indices or coefficients we will examine. STC is the measurement of the capacity of a barrier or material to stop or reduce sound transmission. Nearly any barrier, wall, ceiling, systems furniture panel or floor can be tested and have an STC rating assigned to it. A rating of 45 to 50 STC is pretty high, and 60 STC is probably better than one can hope to achieve in an office space.

(NRC) Noise Reduction Coefficient

This is the second of the two more important gauges. In order to visualize the difference between noise reduction and transmission loss, envision a room built of solid concrete walls, ceiling and floor 8 feet +/- thick. We will assume the room to have an STC as high as possible, and that there is no absorption. In the room is a large bell which is struck one time with a hammer. The bell vibrates, giving off a loud sound which continues as long as the bell vibrates (duration). The sound energy being produced by the bell reverberates and reflects throughout the room, none escapes. As the sound waves strike the surfaces, they are reflected back into and add to the sound field. Eventually, the sound decays through loss of energy by being converted into heat, and ultimately ceases.

Next, envision insulating the inside surfaces of the concrete room with a noise reduction (NR) or sound-absorbing material such as fiberglass batting. When the bell is struck, the sound is perceived to be much less loud and seems to decay faster. In fact, that is what would occur. The sound-absorbing fiberglass batting acts as a noise reduction (NR) material and absorbs the sound energy. Because the batting is absorbent (limp) there is less reverberation and less reflection, therefore less is being added to the sound field. The sound was reduced by absorption into limp, low-reflective surfaces.

Noise reduction is the reduction in sound pressure levels between two points—the point of the sound emission and the point of its perception. The *noise reduction coefficient* (NRC) is a series of mathematical evaluations of a particular noise-reducing material or fabrication at each of the four middle frequencies, 250, 500, 1,000, and 2,000 Hz, averaged together.

The noise reduction coefficient is the measurement of the capacity of a material to absorb or reduce sound energy. Technically, NRC is the ratio of the amount of sound energy absorbed by the material, compared to the total energy reaching the material. A NRC should be 0.5 or better to be qualified as *sound absorptive*. Think of NRC as a percentage of perfect, and perfect is 1.

Keep in mind that a sound absorbing material not only absorbs sound, it also reduces reflection and reverberation in the sound field.

A doubling of the sound absorbing material will yield a reduction of about 3 dB, and a reduction of 10 dB will seem half as loud to our ears.

Once in a while, a product might be advertised or specified with an NRC of, say, 1.17, even though a perfect NRC is 1.00. This is due to the theoretical ability of the particular product to absorb sound through its edges or perhaps through its back.

IIC Impact Isolation Class

Impact Isolation Class originated as a specific measurement of transmission loss in floors and is used primarily in residential construction to isolate impact sounds from one residential unit from its neighbor. It considers a source of sound which could be easily overlooked in construction designs.

IIC measures the transmission loss capacity of a sound barrier to a sound generated by a tapping or impact source such as hard-heeled shoes on the floor. IIC calculations are usually performed in the laboratory, not in the field. An IIC of 100 is perfect.

Impact sounds

Impact noises are transmitted by the building structure itself and become airborne, as they are emitted by the structure. Once an impact sound is generated in the structure, it can travel (or flank) long distances, through any junctions of floors, walls, ceilings, pipes, ducts, etc. adjoining the surface receiving the impact, unless an isolation or attenuation material is designed into the structure to prevent transmission, or the structure is of such a homogeneous mass, such as solid, cast-in-place concrete, that little transmission takes place.

The best cure for impact-generated noise is to isolate it as close to its source as possible. Carpeting is a good floor covering for sound isolation. Carpeting with padding is better. The use of double stud walls, resilient channels for gypsum ceilings, and resilient caulking between gypsum corner joints are all excellent approaches.

Sources of Sound

There are three sources of sound which can be classified as typical office noises. They are exterior ambient, interior ambient and interior operational. Each of these has to be dealt with uniquely and on an as-needed basis. Treatment will be explored after first examining the sources.

Exterior Ambient

Exterior ambient noises are generated outside of the building itself, such as:

- Low flying aircraft; taking off and landing.
- Railroads and light rail systems, particularly subways.
- Highway rumble.
- Street noises.
- Sirens.

Interior Ambient

Interior ambient sounds are generated inside the building — part of which are normally building-related and part of which are operationally oriented. Not all of these sounds may be a objectionable. In fact, a few may be advantageous as background noise, sometimes referred to as *random noise*. The most common interior building related noises are mechanical room noises such as:

- Fan rumbling; these noises are particularly difficult to deal with, as they are usually low frequency sounds.
- Pumps.
- Compressors.
- Water lines; usually these are related to pumps.
- Drain lines.
- Air duct and diffuser sounds; these are often advantageous as background sound, or random noise.
- Elevators.
- Toilet rooms, jet-flush sounds.

Interior ambient sounds not directly related to the building include:
- Copy and fax machine operations
- Vending machines
- Impact printers
- Typewriters
- Telephone ringers.
- Paging systems.

Interior Operational
Interior operational sounds are generated by workers being present and performing work functions.

- Conversation – person-to-person dialogues
- Conversation – telephone
- Conversation – reception areas
- Conversation – Extemporaneous congregating areas
- Voice-activated computer instruction
- Audio-visual and training rooms
- Food service areas
- Clinics

What Kinds of Solutions?

Design
Before dealing directly with the specific problem sounds, it would be helpful to examine the various techniques available for dealing with noise. Some solutions are achieved through sound-preventative design. Other solutions are obtained through a design-curative process, such as adding high NRC materials to walls or ceilings; and some with a cover-up or cancellation treatment.

To avoid airport or rail noise, we should of course select our office away from airports or rail lines. If such a selection is impractical, inquire of the landlord what efforts he has taken to deal with these problems before attempting to do so yourself. If, during the inspection of possible sites and buildings, a site is found to be next to an airport or on a flight path, or next to a major highway, see if an alternate building might not be more acceptable. Once the building selection has been made, the designer or architect should be reminded that exterior ambient sound emission control may be necessary.

We will not try to deal with all of the alternate fabrications available to solve specific noise control problems. However, there follows some of the more useful treatments or solutions.

High STC Design
These solutions are intended to prevent sound from penetrating from one space to another or from one point to another by stopping it at a barrier.

This can be done through such methods as:

- Using mass (additional layers of gypboard for example)
- Eliminating flanking paths (sound traveling around corners)
- Separating transmission joints with caulking
- Using isolation channels under gypboard
- Creating dead air spaces within walls or ceilings
- Using solid septa (solid dividers or separators) in systems furniture panels
- Using sound-insulating materials within walls or ceilings
- Separating wall components through the use of staggered studs and separate plates, or isolation channels
- Using seals at doors, electrical and telephone outlet boxes
- Using high STC ceiling panels or tiles
- Using solid-core or filled steel doors
- Using sound baffles in plenums
- Using double-glazing at exterior windows
- Tilting interior glazing to reflect the noise upward toward the ceiling
- Adding lightweight concrete to structural floors
- Using heavy pile carpeting or carpet pad
- Using isolation devices, such as springs for mechanical equipment
- Proper alignment of motors on their bases
- Balancing air-handling fans
- Isolating supply and return water and refrigerant lines and air-handling ducts from the building structure
- Using isolation collars in air-handling ducts
- Floating or isolated mechanical equipment mounting slabs
- Maximizing static efficiencies of supply and return fans (running fans at peak loads)
- Lining air-supply and return ducts with acoustical materials to prevent transmission
- Careful placement of lighting fixtures in relationship to normal speech frequencies and direction

High NRC Design

This design technique contemplates the reduction of sound within a given area through the use of sound-absorbing materials. Some solutions include:

- Designing rooms as close to square as possible to maximize the floor-to-wall surface ratio and reduce reverberation
- Keeping ceilings as high as possible, especially in narrow spaces
- Eliminating or reducing windows or other highly sound-reflective surfaces
- Using high sound-absorbing (NRC) ceilings about 65 and an STC of 25 to 26, or greater, especially in the areas of high speech occurrence
- Using parabolic-louvered lighting fixtures instead of flat lens diffusers
- Using high NRC materials at flanking paths
- Isolating individual workstations by as much distance as possible. A 10'-6" distance can produce a 10 point drop in dB. 8' is about what designers are currently achieving
- Designing workstation positions to direct normal speech patterns away from adjacent neighbors
- Using sound-absorbing panels between workstations. These panels should:
  - have an NRC of .75, for low-level speech privacy
  - have an NRC of .85 for high-level speech privacy
  - extend to the floor
  - be not less than 60 to 66 inches high

- have an STC of 25 to 26 or greater
- be about 8 feet wide
- Using sound absorbing wall surfaces or sound absorbing wall baffles in critical areas
- Using sound baffles and resonators in ceiling plenums
- Using heavy pile carpeting laid snugly up to the walls
- Enclosing high noise generating equipment such as impact printers in sound absorbing areas or covers

Sound Removal

The two previous sections have touched on the majority of acoustical control techniques. The coordinator of the construction project is ultimately responsible for judging the adequacy of the acoustical treatment provided in the new facility. One of the best ways to control undesirable noise is to eliminate it before it becomes a problem. Much of this can be done by paying careful attention to the very basics of good project design and control. Consider those solutions in High STC and NRC design, outlined above. Don't let the noise in.

Locate noisy departments or groups where they will not bother others. Separate noisy groups from sensitive areas, i.e., marketing and accounting.

Blocking

Blocking is stopping sound from transmitting from one area to another. If you cannot eliminate it at the source or if it cannot be isolated, block it.

Review the various options outlined in High STC design and employ them as close to the source of the noise as possible. The farther a sound transmits, especially through solid materials or ducts, the more it radiates and the more costly and difficult it will become to control.

The use of a couple of additional layers of gypboard and some fiberglass insulation in a wall cavity is not nearly as costly as one might think. After all, the studding is already in the wall cost. The fact that you have *blocked* the sound within the space where it is generated will most likely be the least costly and most efficient solution.

Absorbing

Many of the more commonly used methods of sound absorption are reviewed in the section on High NRC design. Laboratory tests also indicate that a thicker, low-density material is more efficient, especially at lower frequencies, than a thinner high-density material. This distinction is important because the frequencies which were most affected by the thicker low-density materials were the ones corresponding to normal speech.

The net effect is that for any level of speech privacy some, form of NRC control will most likely be needed.

Speech Privacy

A word of caution at this point: Do not get carried away in trying to create "The All Quiet Space" or assure perfect speech privacy. A moderate level of ambient sound is beneficial to obscure normal speech or conversations. A very acceptable level of ambient office sound would be in the 30 to 40 dB range. Much higher than this, and people begin to react to the constant sound as an irritation. Most people do not need total speech privacy. The majority of company business does not involve subject material so confidential that our neighbors in the adjoining workstations must be excluded.

However, attempt to achieve the best solution possible, which would be about 20 dBA at the point of perception. A normal, or conversational voice level, will be about 60 dBA. An elevated voice level will be about 70 dBA. If the ceiling has an NRC of 70 or greater and our 66" high partition has a 35 STC and an NRC of 40, the resulting sound field will be about 35 dBA. About the best we can effectively mask will be about 30 dBA without creating new problems by generating too much masking sound. It is apparent that total speech privacy may have to be achieved through something other than open-plan or systems furniture. It may have to be dealt with through design and planning techniques or other methods.

Masking

Since it is not feasible to provide optimal STC nor NRC levels throughout a facility, at least within normal budget constraints, the use of sound masking should be examined. This system is sometimes called *random noise* or *white sound*. Often, in the past, the air movement sounds generated from HVAC systems supplied a form of *random noise*. It often did the job, however annoying.

Another form of cheap random noise was the use of background music, much of which began to irritate the listener within a very short time. Employees began to bring their own earphone-equipped radios to blot out the objectionable canned background music.

The problem in relying on HVAC noise or background music is that the frequencies generated by these sources are only a segment of those needed for an effective masking system. In addition, the manufacturers of HVAC equipment have been working hard, and successfully, to reduce HVAC noise.

Today's masking sound systems are easily installed and produce a background sound much like a soft breeze. They can be tuned to the specific frequencies needed in a given area to produce a non-fatiguing sound that will mask most speech and provide a desired level of speech privacy. They can be adjusted or tuned in zones as required to offset noisy or not-so-noisy areas.

Cancellation

Sound cancellation or *destructive interference* is a newer form of sound elimination. These systems consist of a microphone located in an area where an objectionable noise is transmitted. The microphone picks up the sound and sends the signal to a microprocessor which analyzes the sound and discerns the various sound waves. The processor then sends a mirror-image signal to a speaker which emits sound exactly opposite in phase to that of the propagating noise. The inverse sound waves tend to cancel the original sound waves, either eliminating the noise or substantially reducing it.

These systems are quite new and are used particularly in HVAC ducts where other forms of acoustical control are physically difficult or present inadequate or undesirable side effects, such as increasing the air static pressure and thereby increasing energy costs. Passive acoustical materials such as fiberglass or foam are effective at higher frequencies, but are most ineffective at the lower frequencies. It is the lower frequencies which are more often generated by large air-distribution fans.

Dealing with sound in an office space is a not a precise science. It is relatively easy to deal with some mechanically generated sounds such as the sounds from an air diffuser, because it is usually from one source, in one location, and at constant and determinable frequencies and dBs. Dealing with the sounds (noises) people make is another matter. We move from place to place. We move our heads to project our voices from one direction to another. We raise and lower our voices. We utilize noise-generating devices such as telephones, fax machines, impact printers, etc., which are also moved from place to place and emit differing frequencies and dBs. Controlling the noises people make is much more complex and difficult to achieve. An understanding of sound, how it is transmitted and what can be done to absorb or block it is what we have attempted to cover in this section. No single solution will work for all situations. Each problem or condition will have to be dealt with uniquely. However, given the degree of attention being paid to acoustics by the various manufacturers of interior products, the company and its professional consultants should be able to achieve and sustain quality acoustical solutions.

Chapter 18

COLOR, PAINT AND WALL COVERINGS

Introduction This chapter discusses color, which is essentially a design element, and also paints and wall coverings, which are two of the most prominent means by which color is introduced into a facility. Carpeting and the fabrics used for seating and systems furniture panels are two of the other most common means of color application.

There is in this chapter a great deal of material which is quite technical and may seem unnecessarily detailed. Its purpose is to review the many aspects of color selection and to provide sufficient information to facilitate working with the company's consultants. Many of the most sophisticated consultants have great "taste" in color, yet lack a fundamental knowledge of the physiological and psychological effects of color on people in the specific work spaces which constitute a typical facility.

Likewise, specification of a wall paint by stating that it should be a "flat latex paint, applied in three coats" is apt to produce a flat wall finish which may not be at all suitable to its particular environment, traffic and maintenance needs.

Color Any discussion of color should begin with a review of the basic definitions of the major aspects of color — Hue, Brightness (Brilliance), and Saturation. See Figure 18.1.

Hue
Hue describes a color as red, yellow, green or blue, or an intermediate between any of these. We can see about 24 full-intensity hues.

Brightness (Brilliance)
Is a measure of lightness or darkness:
—First, from white to black, an achromatic color (value),
—Next, for any chromatic color.

Saturation
Measures the vividness of a color using a medium gray, between white and black, as the center. It is a measure of the color's purity. Colors graded above the medium gray are *tints*; those below medium gray are *shades*.

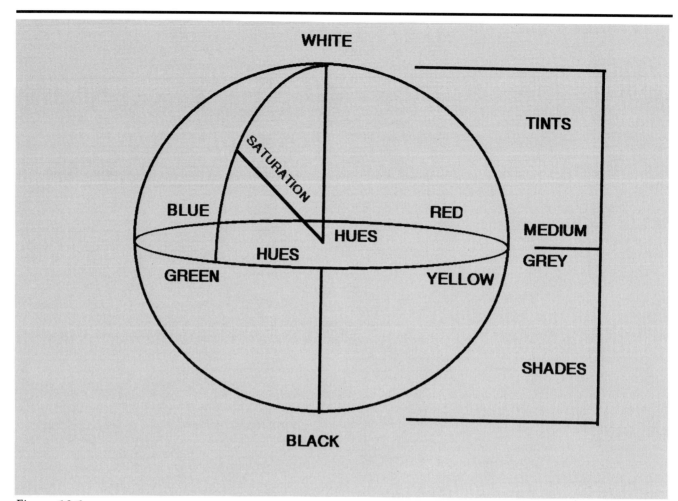

Figure 18.1

Physiology of Color

Students of color seem to have few areas of agreement as to what physiological effects color has on people. These areas of agreement are by no means universal, particularly between the physiological and psychological. There follow some conclusions based upon the most logical and universal conclusions within today's body of knowledge.

Some colors seem to have some physiological effects on the body. In particular, red seems to induce a more rapid heart rate, an increase in blood pressure, an increase in adrenal output, possible irritation, and possibly depression.

Blue, on the other hand, has a nearly opposite effect, tending to relaxation. Violet-blue and turquoise seem to produce a drop in blood pressure, heart rate, and adrenal output and provide a possibly calming effect.

Color Balancing

There seems to be a physiological need for a balance in the amount of warm versus cool colors that surround us. Too much warm color seems oppressive and tiring, while too much cool color seems to make us feel cold and isolated. Our bodies seem to demand an adjustment to too much of any one color scheme. The effects of color shift may be seen by looking hard at a dot of red for a minute or so and then quickly transferring your focus to a plain white sheet. You should quickly see a blue-green dot of equal size.

Reflexive Response

From a basic study of light rays, we know that the warm spectrums – reds, oranges, yellows – are of a longer wave length than the cooler blues, violets and greens. The human eye sees different wave lengths in different ways. Long (warm) wave lengths appear closer to us, while short (cool) wave lengths appear to move away, or appear further away from us. The same effects are true of brightness. The brighter colors seem to move away from us, while the darker colors appear to move towards us.

It is possible to derive some practical applications from these reflexive responses. Surrounding a bright hot color with a dark background should make it pop out at the viewer. Using a warm color at the end of a room may help draw the room closer to us and give it a more intimate feel. On the other hand, the use of a cool bright color should make the end of the room appear to move further away and enlarge its feel.

As people grow older, the lens of the eye begins to turn yellow. This condition effectively filters out some of the subtle effects of lighter tints, converting them to grays. This would suggest a lack of sensitivity to and lack of appreciation for lighter colors and pastels by older workers.

Rod/Cone Shift

As an environment becomes darker, that is, having fewer foot candles of light, the eye shifts from predominantly cone-oriented to rod-oriented. The rod sensory bodies of the eye are light-oriented with less color perceptivity. They are short-wave-length-oriented and are capable of seeing blues much more easily than the warmer, long-wave-length colors such as reds. The cone sensors are more color-oriented and capture warmer, brighter, long-wave light emissions.

Psychology of Color

The previous portions of this chapter discussed physiological or reflexive responses to color. We prefer to deal with all responses to color as psychological, acknowledging the physiological functions, but attributing the majority of the responses to the psychological makeup of the individuals involved. In other words, our approach is to view most color response from the point of view of the employees for whom the work environment is being created.

Environmental Conditioning

There are certain universal environments in which we all live, and which we all see or experience. Clear blue skies have a calming, cooling, tranquil effect on almost all of us. Gray skies can have an opposite effect, often depressing, and if the skies are heavily laden with clouds, the gray skies can generate a quite depressed psychological state. The chart in Figure 18.2 lists the major colors of the environment, and some of the psychological effects they can create.

Sociological (Cultural) Conditioning

Certain sociological or cultural conditions affect our responses to color:

Age

Younger people generally prefer brighter, more vivid colors, with strong preferences for the primary colors. As people grow older and have more life experiences, their color tastes appear to change to more subtle choices. As old age approaches, these tastes appear to again change to an interest in the more intense colors, probably due to the change in the physiological color of the eye's lens.

The Psychological Effects of Color

Blue
Sky, oceans and lakes ... soothing, calming, feeling of well being.

Green
Fields of grass, cool forests ... calming, cooling.

Yellow
Flowers, warm sunshine ... happiness, cheerfulness, optimism, warmth, friendliness.

Orange
Flowers, warm sunshine ... warmth, firesides, an appetite stimulant.

Red
Fire, heat, blood ... depression, fear, warmth, anxiety.

Brown
Earth, tree trunks ... strength, warmth, calming, fatigue.

Black
Night ... confinement, fear, depression, calming.

Figure 18.2

Education

There seems to be a strong correlation between our levels of formal education and our tastes in color. The more educated a person, the more likely the choices will tend away from the intense colors, and toward the lighter tones and tints.

Economics

There seems to be some evidence to indicate that those who spend their lives at lower income levels often seek out more vivid colors, regardless of hue. Conversely, middle-to-higher income groups seem to accept a much broader palette of hues, brightness and saturation. Within this same socioeconomic group, people who are from impoverished environments will seem to seek out colors that are purer, warmer, and have a higher visibility. They may have a strong affinity for bright greens, which are often absent from large, inner-city environments.

Emotional Outlets and Responses

The same socioeconomic studies just mentioned indicate that individuals with many emotional outlets, such as dining out, vacations, theater, etc., will prefer lighter tones and tints, and colors with less vibrancy or brightness. On the other hand, individuals with limited emotional outlets seem to be attracted to and function better in color schemes having greater variety and intensity.

Middle of the Road

In selecting the color palette for any given area, keep in mind that staying too close to the middle of the road can also be unsuccessful. Schemes that are too neutral can have an irritating and fatiguing effect on all cultural groups because of the lack of accent colors or undertones to give the scheme substance.

The color schemes selected should be sensitive to the various employee groups working in given portions of a facility, as well as to the work functions being carried out within a given area. Obviously, the scheme selected for a mail room will not be the same as one selected for a board room; nor should the scheme selected for a high-stress telephone reservation system department be the same as that selected for a group of researchers or actuaries.

It is of interest to note that of the four major criteria for a purchase selection – function, form, color and price, nine of ten people will place color first.

Paint

The use of paints to enhance wall surfaces dates back many centuries. Evidence has been found that the early Egyptians, Greeks and Romans all decorated their palaces and temples with paint. Medieval cathedrals and churches still have evidence of polychromed walls and design elements. It is therefore not surprising that we should be concerned with paint.

Not only does paint provide inexpensive options for varying color scheme environments, it also provides us with various protective coatings on walls, ceilings, floors and other surfaces. Paint is by far the most versatile coating available.

In recent years the paint industry has broadened its product base to include paints tailored to provide protection for a multitude of environments and surfaces. Many manufacturers produce well over 100 different formulations, each intended for a different use and condition.

While all of this has been of benefit to the user, there are now developing problems. Most of these original formulations have contained some form of VOC, (*volatile organic compounds*), such as mineral spirits, etc.

When selecting paints, be particularly aware of types, uses and availability of the many paint products available. The very nature of paint has made it of great concern to our environment.

The Nature of Paint

Most paints are made up of three principle parts or components:

Pigments
Usually powdered metals, ceramics or minerals blended together to provide thickness, color and sometimes sacrificial materials.

Binders
These are liquids which harden and form the protective film along with the pigments. They are generally made up of resins from such natural plant sources as linseed and tung oils and other fatty acids; alkyds from the glycol and alcohol families; epoxy resins; polyester resins; silicone resins; vinyl resins; acrylic resins; and latexes, now mostly synthetic rubbers.

Solvents or Water
These are other liquids which, together with the binders, form the paint *vehicle*. To these solvents and binders are added driers, oxidizers, antifoams, antisettling agents, various preservatives, bactericides, insecticides, corrosion and ultraviolet inhibitors, catalysts, and a host of other agents.

Problems and Trends

It is the high-gloss, oil-based or so-called solvent-borne paints that contain the most toxic chemicals such as naptha, xylene, and toulene. These are used to provide spreadability and curing characteristics. When removed from the paint, substitute ingredients must be found which are non-toxic and will still provide an adequate performance substitute.

An example of a VOC problem is lacquer. It produces a finish highly desired for its depth of color and high gloss, particularly for furniture. A gallon of lacquer weighs about eight pounds. Of that eight pounds, six pounds is solvent, all of which evaporates into the air as a contaminant. So far, the paint industry has no equal substitute for lacquer.

A number of states have passed legislation restricting the grams per liter of VOCs permitted in specific urban areas. One state, New Jersey, has passed statewide legislation governing the amount of VOS (Volatile Organic Substances) permitted. A large number of states have similar legislation pending. Most of the VOCs permitted vary from state to state. While this form of legislation is highly desirable from an environmental standpoint, it is creating problems within the paint manufacturing industry by creating enormous inventory problems, as well as problems in providing suitable substitute products with equal performance characteristics.

These problems either have or will become the company's problems. The paint used recently may no longer be available. There may or may not be a suitable and comparable substitute.

Solutions and Trends

The paint manufacturing industry is trying to get Federal legislation created to set uniform VOC standards in order to avoid uncontrollable inventory and manufacturing problems. Any sort of Federal Standards appear to be some time into the future.

The industry is also trying to create new paints and paint systems, such as the high-solid paints. Some of these are pure solids, totally without a vehicle, and require high temperature spray systems for application. If high-solid paints are formulated with a vehicle, they are very slow drying.

The paint industry is also creating enhanced formulations (low VOC) water-based vehicles, of which there are two types:

Water-Reducible
This formulation is shipped dry. The water is added at the site, with the resins conditioned to accept the water. Currently, the product is extremely difficult to mix, is subject to rapid settling, and may adversely react when in the can.

Water-Borne
These are the well-known latex paints in which the water is manufactured into the resins. These paints still have poor high-gloss characteristics, poor color resistance, poor adhesion, poor blocking, and often poor hiding. On the other hand, they are usually easy to apply and emit mostly water vapor when drying.

Characteristics of Paint

On a construction project it is common to specify and use ten to twelve different types of paints and varnishes. End product concerns are usually hiding, durability, adhesion, resistance to soiling, washability and VOC content. You may have some special concerns relating to flexibility, impact, abrasion, and corrosion resistance.

Figure 18.3 illustrates how paints might look in a cross-section when dry, and what the components might be for the three most typical types of paints to specify: flat, egg-shell, and enamel. The components listed are typical and for illustration purposes only. They should not be construed as specifications.

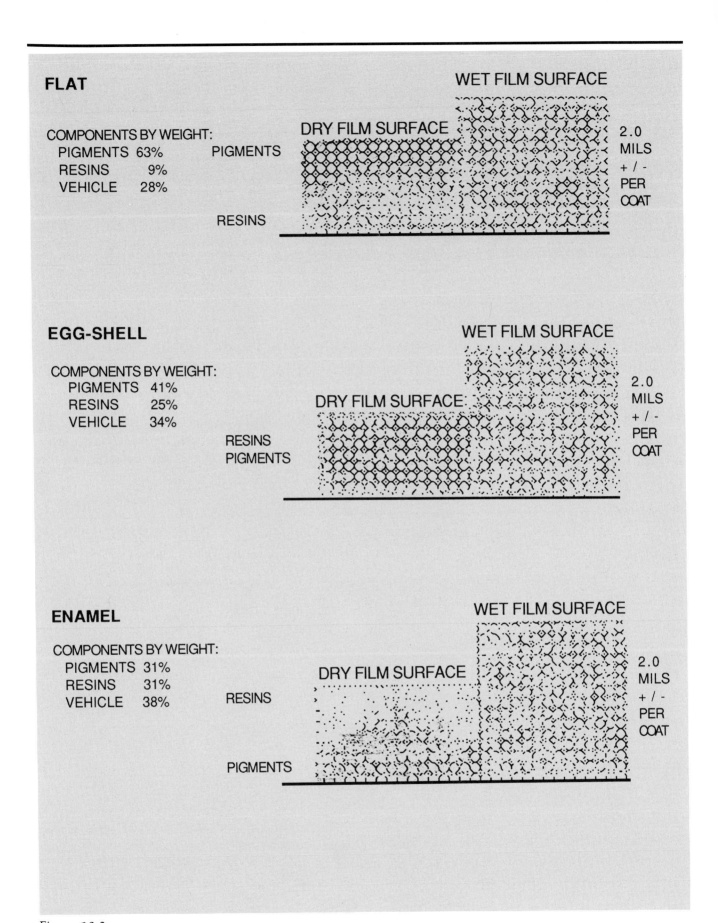

FLAT

COMPONENTS BY WEIGHT:
PIGMENTS 63%
RESINS 9%
VEHICLE 28%

PIGMENTS

RESINS

WET FILM SURFACE

DRY FILM SURFACE

2.0 MILS +/- PER COAT

EGG-SHELL

COMPONENTS BY WEIGHT:
PIGMENTS 41%
RESINS 25%
VEHICLE 34%

RESINS
PIGMENTS

WET FILM SURFACE

DRY FILM SURFACE

2.0 MILS +/- PER COAT

ENAMEL

COMPONENTS BY WEIGHT:
PIGMENTS 31%
RESINS 31%
VEHICLE 38%

RESINS

PIGMENTS

WET FILM SURFACE

DRY FILM SURFACE

2.0 MILS +/- PER COAT

Figure 18.3

Definitions

Adhesion

Adhesion is the ability of a coating to stick to its undersurface. Some paints have inherently good or poor adhesion characteristics. All can adhere fairly well if the undersurface is properly prepared.

Blocking

Blocking is the ability of a coating to dry completely. Some coatings contain plasticizers which prevent a paint from ever completely drying.

Hiding

Hiding is the ability of a single application of paint (the amount of paint normally applied in one coat by brush) to cover its underlying surface, without letting any of the underlying color show through. Some paints, such as vivid reds and bright yellows, may require additional coats to achieve the specified intensity.

Gloss

Gloss is a measurement of light reflectance of a paint surface. A beam of light is directed at a painted surface at 60 degrees (for most surfaces) and at 85 degrees (for flat surfaces). The amount of light reflected off the surface at the same angle, measured by a photoelectric cell, is the measurement of gloss. Perfect gloss (a mirror) = 100% reflectance. Dead flat = 0% reflectance. Note the overlap of ranges by type in Figure 18.4.

Color Change

Color change is the inclination of a coating to change color throughout its life. Some colors are particularly sensitive to color change. Most color change is caused by sunlight (ultraviolet rays). Obviously outside paints will erode, or chalk, as well as fade due to wet weather. Interior surfaces will change color less than those outdoors, but in direct sunlight, they can still suffer considerable fading.

Darker colors, along with reds and yellows, are more prone to fading. Color change is not entirely due to ultraviolet exposure. Some alkyd whites will turn yellow when *not* exposed to light. High gloss and eggshell alkyds will begin to lose their gloss in sunlight. When this occurs, the color change, usually fading, will be quite rapid.

	Paint Reflectance	
Type	**Range**	**Test (ASTM-D-523)**
Flat	Below 15%	85 Degrees
Eggshell	5-20	60 Degrees
Satin	15-35	60 Degrees
Semi-Gloss	30-65	60 Degrees
Gloss	Over 65%	60 Degrees

Figure 18.4

Other Definitions

There are a number of definitions which mostly describe the physical properties of a coating, such as *abrasion resistance, pencil hardness, flexibility, dry heat resistance, wet heat resistance, moisture and condensation resistance, direct impact resistance, salt fog resistance, thermal shock resistance,* and *scrub resistance.*

As the specifications become more stringent, seek help from the customer service department of a major manufacturer. Use one of the painting systems guides mentioned in the following text for ordinary needs, and check with manufacturers' customer service sources if questions arise.

Maintenance

Paint maintenance in an aesthetic environment is nearly a no-win situation. The paints which will have the best scrubability characteristic, such as gloss enamels, will produce the greatest amount of glare (reflectance) and be one of the most difficult to touch up. The more a gloss-type paint is washed, the more gloss surface is worn away, which leaves a quite weak pigment layer.

Flat paints, because most of the pigments are at the paint surface, produce the least glare. This is a desirable characteristic in an office. However, flat paints are the most sensitive to cleaning. As they are scrubbed or even damp-wiped, they will lose some of their surface pigments and *burnish* or take on a polished look. In fact, the cleaning action is a form of polishing.

Eggshell paints produce more glare than flat paints, but not a great deal more. Eggshell paint has the advantage of having more resins intermixed with the pigments. This provides the paint with a harder surface, one less apt to polish when cleaned.

Specifying Paint

There are usually ten-to-twelve different paints that need to be specified for the average facility. The seven elements in the paint specifications will be discussed in this section. The specifications for each of the ten-to-twelve different paints are unique. Rather than writing ten or twelve different specifications, we will review each of the seven elements of the specification and direct you to the product description or specification publication published by all major paint manufacturers. This publication may have a title similar to:

- *Painting Systems Guide*
- *Painting and Coatings for Specifiers*
- *Painting and Coating Specifications*

A telephone call to the manufacturer's local sales office or warehouse should be sufficient to secure that manufacturer's Specifications Guide and probably a set of their color chips.

How to Specify Paint

Select three or four of the major manufacturers who are active and have factory representatives servicing the area. Often a major national paint manufacturer is not active in certain markets, while other manufacturers which are more local or regional may nearly dominate a market.

1. Secure their product description (specifications) publications as described above.
2. Select the paints by function from the manufacturer's product description.

3. Prepare the project's paint specifications using the specification outline contained in this section and the manufacturer's specifications. Do not attempt to shortcut this process or be talked into using a manufacturer's product which does not fit the need. Do not permit a painting contractor (applicator) to talk you into a simplified specification or into using an alternate product which you have not reviewed and approved on the basis that "It's just as good as your selection. I've used it for years." Most painting contractors have favorite suppliers and manufacturers they will wish to use. It's your facility's paint job.

In preparing specifications, have in mind that the specification must fit the surface condition and type of surface, the environment, the conditions at the time of application, the method of application, the ultimate function of the paint (besides color) and the desired life of the coating. Washing a previously painted wall with a 10 percent muriatic acid solution as a preparation would hardly be an appropriate preparation. Washing a concrete floor with a 10 percent muriatic acid solution might be most appropriate.

Seven Elements of Paint Specification

1. *Surface Preparation*
   Over 75 percent of coating failures are the result of improper surface preparation or condition at the time of application. The surface must be sound and dry; clean; and free of oils, waxes, and loose rust (some metals must be white, i.e., completely free of any corrosion), scale, dust or loose paint or other materials.

   One manufacturer lists six conditions and four different methods, one with ten separate steps for the preparation of poured concrete surfaces.

2. *Product Specification*
   Describe the manufacturers' products by their unique product descriptions and product numbers, such as Alkyd Semi-Gloss Enamel, Number B34W200. List acceptable alternates in the same manner. Do not accept substitutes without researching them and without a compelling reason to deviate from your already approved selections.

3. *Method of Application*
   Specify how each coat is to be applied. If by roller, what nap (short, long, etc.). If the paint is to be applied by sprayer or brush, indicate which is needed. Think about what is being painted and where it is. There may be equipment nearby which could be damaged by spraying. If some items require spray-painting, state which type of sprayer: hot, airless, conventional, etc.

4. *Workmanship*
   Describe the work process to be followed. If a prime coat is to applied within a certain time limit after surface preparation, say so. If a second coat is be applied only after 48 hours of the application of a first coat and then at a required temperature and humidity range, say so. Do not assume that the contractor knows all of the appropriate specification data.

5. *Label Directions*
   This is probably the least observed and yet most valuable set of directions available for the applying contractor. The label directions will describe the unique conditions and requirements for the specific paint being applied. Label directions will describe relative humidity (RH) requirements, thinning instructions, maximum and minimum temperatures, time periods before recoating, and drying times. Your

specification should state that the contractor shall read and follow the label instructions. Be sure to look at the cans. Do not accept paint brought to the site in some sort of generic mixing pail, with the painters claiming, "We mixed three or four gallons together out in the truck."

6. *Film Thickness*

It is the final DFT (dry film thickness) that will determine the quality of the paint job. The desirable range is anywhere from three to 5 mils. One mil is one-thousandth of an inch (.001). This should always be a process of building up the DFT by multiple coats. No two coats will be applied exactly the same. The missed area in the first coat will be cured in the second or third coat. Clearly state what DFT is expected, and how many coats are expected to achieve the specified DFT.

The can label will state the volume of solids per gallon in percent of solids to thinner. It should also state the mil thickness when wet at a spreading rate of so many square feet per gallon. To determine the DFT, multiply the volume solids percentage times the mil thickness when wet, applied at the recommended spreading rate.

Example:

Label Data
Volume Solids: 49%
Recommended Spreading Rate (Coverage): 425 sq. ft./gal.
@ 3.75 mils wet, 1.84 mils dry
3.75 mills x 49% = 1.84 mils
If applied at the Recommended Spreading Rate of 425 sq. ft./gal.

7. *Inspection*

The project specification should clearly state that all surfaces may be inspected by a named individual, such as the interior designer, architect or project manager, before the next coat can be applied. If the surface is not acceptable, some form of corrective action, such as touch-up, sanding, wiping, washing, or more drying time, must occur.

Preparing a Specification

If the company is moving into a new, leased space, chances are that the landlord, his architect or contractor will be writing the paint specification. The company or its interior designer or architect will make the color selection. During the discussions with the landlord's representative, request that the company be permitted to specify the paints and provide input into the preparation of the specifications.

Review the prepared Finish Plan and Finish Schedule which will show the color for each wall surface with special notes for cabinets, details and the like. Select the specific paints desired for each surface, indicating the first choice selection and at least two alternates.

The landlord will handle all of the painting needs as a part of the company's lease agreement. Often he will have set a certain maximum dollars-per-square foot of wall surface he will pay. In such cases, the company will have to pay all costs over standard, if any. Typically, a landlord's painting allowance covers the cost of a normal painting requirement. However, the company may want to try to be an active participant in the bidding process and retain the right to approve or reject any and all bids received by the landlord.

Work with the landlord to determine who may be designated to provide inspection approval. Try to have the company's representative monitor the approval process and provide comments to the landlord's inspector. *Do not* request that the approval be made by the company's representative. If a problem should develop, the company does not want to have voided any warranties between the landlord and his contractors or between the company and the landlord.

Complete the Special Conditions (See Master Forms Section) in the places indicated.

If the company is contracting directly, include a set of the General Conditions. These are broad and elaborate General Conditions and are suitable for general construction. They will most likely be employed on large installations. They make an excellent outline of problems and conditions of which to be constantly aware. Their inclusion may seem to be a bit of overkill. However, they will provide fair and reasonable ground rules for the job. (See Master Forms Section.)

If the company will not be awarding the contract and the landlord handles the purchasing, it is not necessary for the company to prepare a complete set of bidding documents. It is only necessary to specify the type and color of the paint you selected and attempt to be involved in specifying the application procedures. To do this, remove, copy, and complete the following pages from the Master Forms Section:

Section 1.   General Conditions
No fill-in necessary.
Use only if contracting directly.

Section 2.   Special Conditions
Fill in necessary information where indicated.

Type the company name, and address, and the date in the appropriate sections. Insert the correct page numbers on the Table of Contents page. Prepare a cover sheet.

Include a print of the completed Finish-Plan Drawing of your facility, on which the various colors and wall covering materials are indicated, with the other documents (but not bound together with them). (See the Construction Documents Section for a sample plan.)

Bind all sheets together in the following sequence:

1. Cover sheet
2. Table of contents
3. Section 1
4. Section 2
5. Drawings

Assemble the complete package and submit it to the landlord. He will contract for paint and application in accordance with the arrangements negotiated with him.

Wall Coverings

Wall coverings originated with tapestries dating back to the time of the ancient Egyptians. They were originally carpets and also reversible. Tapestries were hung on walls to reduce drafts, and for better viewing of the scenes and patterns woven into them, and, perhaps unknowingly, to act as an insulation to reduce heat transfer to stone walls. Wallpapers date back to the 16th century. They were hand-blocked, and considered so valuable that they were mounted on boards so they could be moved from room to room. They were essentially meant to beautify. Today, wall coverings in the commercial office serve some of the same functions. Wall coverings can supply a variety of effects unattainable in paints. In particular, they can provide patterns, textures, a high degree of durability, low maintenance, safety and economy.

Because of their cost-effective durability, bacterial and fungus resistance and their ease of cleaning, they are fast becoming the standard for use in food service areas, health clinics and offices. High-traffic areas and restrooms in these areas, in particular, deserve serious consideration.

Types of Wall Coverings

This section defines five basic types of wall coverings, along with some special types designed to provide solutions to special conditions. There are many more — almost as many classifications and definitions as there are designers and manufacturers.

Wallpaper

A material made primarily of paper upon which a design or pattern has been printed. There is often a coating applied to the print side to provide some additional durability.

Vinyl

A sheet of PVC upon which a design, pattern and/or texture has been applied. The PVC sheet is usually backed with a woven or a non-woven structural material, and sometimes has a coating applied to provide additional durability.

Fiber products

Non-woven yarn- or tufted-type products which are laminated to a backing.

Woven

A woven fabric is usually back-coated to render it suitable to function as a wall covering.

Rigid Surfacing

Usually an unbacked, solid, vinyl-acrylic sheet, textured with the color throughout the entire thickness.

Specials

There are a variety of specialty types including wood, metal, plastic laminates, acoustical, etc.

The Physiology and Psychology of Patterns

Wall coverings are available in a nearly infinite variety of patterns and designs. It is obvious that some patterns are more suited to residential use than to office use. What is not so obvious are the rather strong physiological and psychological effects that pattern seems to have on most people. Some of the same concerns reviewed in our discussion of color should be considered when selecting wall coverings.

Not nearly as much information is available on the physiological effects on people of patterns as there is with color. What is available suggests that some patterns indeed have very strong physiological effects on many individuals. Some of these effects are often very powerful and almost always negative. Designers should be extremely careful in their choice of patterns, and with some particular patterns, even more caution should be exercised. Because of a shortage of data, it is difficult to separate the physiological from the psychological effects of patterns. Therefore, the physiological and psychological will be explored together.

We will discuss six different pattern types and their apparent effects on some people. Just as with color, there are no universal responses, although there are significant generalizations.

Stripes

High-contrast vertical stripes are a pattern which should be used with extreme care. Many people have strong negative responses to this pattern. In some instances, actual vertigo and severe headaches, even migraine, can result from looking at strongly contrasted vertical stripes. Many of us have experienced some visual difficulties with escalator treads or vertical blinds. The act of ironing some striped shirting and blouse materials can bother some people. It should be expected that strongly contrasted, striped wall coverings can cause very negative physical responses.

Low-contrast vertical stripes such as low-contrast strings or rope-like patterns have a nearly opposite effect. These patterns appear to produce little *after image* effect or *pattern movement*. Most people appear to respond favorably to this pattern, considering it to be soothing, orderly and subdued.

Checkerboard

Patterns such as plaids and ginghams, particularly those with strong contrasts such as black on white or hot on cool, frequently elicit strong negative responses. There are very high after image and pattern movement effects. Acceptances run from disturbing to uncomfortable.

Graphic Shapes

Circles and diamonds of varying sizes and colors seem to fall into a mid-range of response, and may hold a clue to the differing physiological and physiological responses. There are higher than average complaints of after image and pattern movement. On the other hand, acceptances run from orderly to harmonious to relaxed and refreshing, varying with the amount of contrast in the pattern.

Non-geometric

Patterns with marbleized, sliced stone or water designs seem to lead the list in terms of greatest acceptance with low pattern movement and nearly no after image complaints. They evoke responses such as subdued, relaxed and quiet.

Pictorial

These patterns include such images as the ubiquitous sea shells, murals and antique advertising pages. The visual responses are interesting, again giving us a clue into the physiological/psychological responses. There is very low after image complaint, but an above average pattern movement. The responses to these patterns are abnormally poor, evoking responses from flashy to discordant and tense.

Floral

The floral-type patterns range from large floral scenes to smaller, nearly geometric types of patterns. They have a very high after image effect, especially with the very young and older people. In these two groups there are sometimes "daymares," in which faces seem to appear in the flowers, and which often reappear in the same locations on the wall. Yet these patterns seem to have better than average responses ranging from orderly to positive.

Visual

One of the added benefits of a vinyl wall covering is the visual effect of the material's texture. Unlike paints of equal purpose in cleanability and durability, wall coverings are nearly flat in terms of light reflectivity. This is in part due to their imbedded textures. Flat wall paints polish when washed. Wall coverings do not. Because of their textured surfaces, wall coverings tend to break up the horizontal light rays and reduce wall-generated reflectivity.

Tactile

Because many wall coverings are textured, they have an inherent tactile characteristic. Usually this condition has the effect of reducing the tendency of people to touch the walls.

Colors

The first part of this chapter dealt with color; that information is fully applicable to wall coverings. Because of the differences in manufacturing and inventory maintenance, wall coverings are available in fewer colors than paints. Usually this is not a highly limiting qualification. Those few walls that require intense or deep-tone colors are normally in low-traffic areas and are normally contrast walls, limited in size.

The wall covering industry attempts to be as current as possible in their color pallettes, and many of the wall covering manufacturers are also in the paint or upholstery fabric manufacturing business. As a result, they are fully abreast of the latest in color trends.

Problems and Trends

Market Perception

Unfortunately, there seems to be an automatic predisposition toward paint as the only wall surfacing material to be used. This is due to three basic perceptions.

First, paint is cheaper. Well, maybe. In the next portion of this chapter we have developed a straight-line comparison of vinyl wall covering vs. paint. Certainly in many areas of a facility, paint is the logical and most cost-effective material. In high-traffic areas, give serious thought to vinyl wall covering.

Second, paint is the easier of the two material to apply. By in large this is true, and it has to be applied to doors, door frames and other forms of trim regardless of the final wall finish. However, when it comes to a two-part epoxy system, neither the ease nor cost of application is substantially better.

Third, paint is without question the more versatile product when looking strictly at color. The interior designer has a near infinite variety of colors from which to choose. Perfect matches can be made to carpets and fabrics with paint. On the other hand, vinyl wall coverings and papers have patterns and textures not available in paints.

As the vinyl wall covering industry continues to expand and to develop broader product lines, particularly in color selections, interior designers may begin to look at some of the elements which differentiate paints and wall coverings.

Cost Perception

There is little disagreement that paint has a much lower *first cost* than any vinyl wall covering. If it is your intent to consider repainting as a part of your program, and repainting is in a three-to-four year time frame, you should do a simple life cycle analysis. This is particularly important in areas of high traffic, or where a high degree of cleanliness is desired and painting might reoccur more frequently than every three years.

Using a base cost of $.31/SF for two initial coats of paint, one of primer and one of latex finish coat, and a cost of $.23/SF for a one-coat re-paint every 3 years, we will have a total cost of $1.00/SF at the end of nine years and $1.23/SF at the end of twelve years. Note that this is a straight line projection with no allowance for inflation such as increases in paint costs or painters' labor rates.

Utilizing a type II vinyl wall covering, applied at $1.07/SF initial cost and a $.033/SF washing cost every five years, we will have a total cost at ten years of $1.13/SF. Again, this is a straight line projection with no allowance for the increases in wall washing costs.

A comparison which considers the increasing paint costs and painters' wages over the nine-to-twelve years will produce a crossover of the two methods closer to the eighth or ninth year.

Vinyl Wall Coverings

Vinyl wall coverings are grouped into three major categories. These categories are determined by the wall covering's weight per square yard.

Type I
Type I is normally considered a lightweight material and is usually a paint alternative used in light-traffic areas such as offices, both private and open plan. It can provide a good low-glare surface with reasonably good cleanability. It uses either a scrim or non-woven backing.
Weight: 7 to 13 oz/sq. yd.

Type II
Type II is the middle-weight group. These wall coverings are the most commonly used and provide excellent service in the majority of areas: classrooms, dining rooms, elevator lobbies, hospital wards and rooms, lounges, public corridors, and stairways. They are available in a wide selection of colors, patterns and textures. They use either Onsaburg, non-woven, or sometimes a drill fabric backing.
Weight: 13 to 22 oz/sq. yd.

Type III
Type III is the heavy-weight wall covering. These materials are intended for high-traffic, high-abuse areas subject to heavy traffic and possible contact with vehicles such as gurneys, food service carts, delivery dollies, etc. If these conditions exist and a high degree of cleaning is necessary (such as in a food service area or hospital corridor), these materials probably should have a special coating. Even then, some form of chair rail or bumper guard may be advisable. They are almost always backed with a drill fabric.
Weight: 22 + oz/sq. yd.

Standards
Most vinyl wall coverings come in either 27" or 54" widths, with 54" being the most common.

Most Type I products come in a 60-yard bolt.

Most Type II products come in a 30-yard bolt.

UL ratings are usually available with orders of:

Type I – 1,000 yards;

Type II – 750 yards;

Type III – 500 yards.

Regulations and Specifications
Most all quality wall coverings meet Class A ratings:

ASTM E-84 Flame index of 0-25; and Smoke Development of 0-50

Most manufacturers of vinyl-coated wall fabrics publish information on many test specifications such as tensile strength, tear strength, abrasion resistance, adhesion, color fastness, blocking, crocking, cold crack resistance, heat aging, shrinkage, stain resistance, etc. If the intended application warrants close scrutiny of these qualities, make sure the selected product meets all necessary quality tests. If more information is needed, check with:

Chemical Fabrics & Film Association, Inc. (CFFA)
1230 Keith Building
Cleveland, Ohio 44115
216/241-7333

Most products of the domestic manufacturers who belong to the above association will have received the approval of New York City MEA tests. Most vinyl wall covering products do not appear to qualify for labeling requirements under the *significant risk* regulations promulgated by California Proposition 65.

Characteristics of Wall Coverings

How Wall Coverings Are Produced
Vinyl wall coverings are made by one of two basic methods:

Calendering
Calendering is a process of rolling a compound of vinyl material through a series of hot cylinders (calenders) or rollers which squeeze the material into a film or sheet. This sheet of goods may be used as is, or it may be laminated onto a fabric backing during the final stages of the calendering process. An alternative final stage may involve laminating the vinyl film in a separate machine using an adhesive in lieu of the heat and pressure of the calendering machine.

Plastisol Process
The plastisol process is a direct application method. It begins with a moving base fabric, over which is spread a viscous vinyl, using a stationary knife or scraper. When the appropriate vinyl thickness has been achieved, the vinyl is fused to the fabric at a high temperature.

Backing Materials
There are currently about four different backings in use. The differences between them functionally is only the degree of strength desired by the user.

Scrim
Srim is a lightweight woven fabric used in TYPE I wall coverings. It has moderate strength and weighs between 1.1 and 1.5 ounces per square yard.

Osnaburg
Osnaburg is a moderate-weight woven fabric looking a bit like a coarse piece of sheeting. It is used in TYPE II wall coverings. It is good structural strength, is resistant to tearing, and weighs between 2.0 and 4.0 ounces per square yard.

Drill
Drill is a heavyweight material with a characteristic diagonal third thread. It is structurally very strong and tear-resistant, and is used in the construction of TYPE III wall coverings. It weighs between 2.5 and 4.5 ounces per square yard.

Non-woven
Non-woven backings are cellulose polyester filaments that are felted onto a fabric. They provide structural strength and weigh between 1.0 and 3.5 ounces per square yard. They are used in TYPE I and TYPE II wall coverings.

Special Coatings
There are at least three special coatings which can be applied to most vinyl wall coverings. These special coatings can offer greatly improved cleanability, stain resistance and surfaces with good bacterial and fungal inhibition characteristics.

One of these special coatings, acrylics, does not seem to be quite as effective in durability tests as the other two products. Of the other two systems, one is a film, laminated to the basic wall covering. It is extraordinarily strong and stain-resistant, with superior cleanability. However, it has somewhat of a luster and seems to *bridge* over deeply textured wall coverings. The third system is a liquid coating applied during the manufacturing process. It, too, is extraordinarily strong, only slightly less stain-resistant and cleanable, but because it is applied as a liquid, there is no bridging and virtually no lustre.

The use of special coatings should be considered for health service areas, food service areas, restrooms, and clean rooms, or wherever considerable wall cleaning might be necessary.

How to Specify Wall Coverings

The selection and specification of wall covering does not permit the same latitudes available when specifying paint. The company or its designer will be selecting very specific manufacturers and patterns. It will not be able to permit contractor-quoted alternates. However, you may have to re-select in the event of an out-of-stock material – which is not an unusual event.

For each manufacturer selected and for each product line produced by the manufacturer, secure their product description (specifications) publication.

Prepare your own specification using the manufacturer's specifications. Do not attempt to short-cut this process.

Surface Preparation

The preparation of a newly completed gypboard surface will be different from the preparation of an older plaster surface painted with a high-gloss enamel. The surface must be sound and dry, clean and free from oils, waxes, loose paint, dust or other materials. All holes should be patched and lightly sanded so that the surface is level.

Product Specification

Describe the manufacturers' products by their unique product descriptions and product numbers. If you have pre-selected alternates, list these in order of preference. Do not accept substitutes without doing research and without a compelling reason to deviate from the already approved selections.

Method of Application

Specify how each wall covering is to be applied. In particular, specify which adhesives should be used. Some wall coverings have special application procedures to minimize matching problems and obvious seaming. If a primer coat is necessary, specify what type of primer should be used.

Workmanship

If a prime coat is to be applied to a given wall, say so. If the drying time for the prime coat is 24 hours, say so. Do not assume that the contractor knows all the appropriate specification data. Again, follow the manufacturers' specifications to the letter.

Inspection

Your specification should clearly state that all surfaces must be inspected by a named individual, either the interior designer, architect or project manager, before the wall coverings may be hung. If the surface is not acceptable, a corrective action such as priming, sanding, washing or more drying time must occur.

Preparing a Specification

If you are moving into a new, leased space, it is likely that the landlord, his architect or contractor may (or may not) be writing the wall covering specification. The company or its interior designer or architect will be doing the wall covering selection. At this point in the discussions with the landlord's representative, request permission to specify the wall coverings wanted or to provide input into the preparation of the specifications. After all, the chances are very good that the company will be paying for a large portion of the cost as an above building standard cost.

Work with the landlord to determine who may be designated to provide inspection approval. Try to have the company's representative monitor the approval process and provide comments to the landlord's inspector. As with any work performed by the landlord, *do not* request that the approval be made by your representative. Remember, if a problem should develop, you do not want to have voided any warranties between the landlord and his contractors or between the company and the landlord.

As with any work the company is contracting directly, include a set of the General Conditions. These are broad and elaborate General Conditions and are suitable for general construction. You will most likely employ them on larger installations. These General Conditions make an excellent outline of problems and conditions of which you should be constantly aware. They will provide fair and reasonable ground rules for your job. (See the Master Forms Section.)

Complete the Special Conditions in the places indicated. (See Master Forms Section.)

Review the Finish Plan and Finish Schedule which will show the material for each wall surface with special notes. (See the Construction Documents Section.)

> Section 1.   General Conditions
> No fill-in necessary.
> Use only if you are contracting directly.
>
> Section 2.   Special Conditions
> Fill in necessary information where indicated.

Type your company name, address, and the date in the appropriate places. Insert the correct page numbers on the Table of Contents page. Prepare a cover sheet.

Include a print of the completed Finish Plan Drawing of the new facility, on which the various colors and wall covering materials are indicated, with the other documents, but not bound together with them. (See Construction Documents Section for a sample plan.)

Bind all sheets together in the following sequence:

1. Cover sheet
2. Table of Contents
3. Section 1
4. Section 2
5. Drawings

Assemble the complete package and submit it to the landlord. He will purchase the wall coverings and coordinate the installation in accordance with the arrangements negotiated with him.

INDEX